Math Skills Workout

Grade 3

About This Book

Use this super resource—*Math Skills Workout Grade 3*—to help get your youngsters' math skills in tip-top shape! Inside you'll find just what you need to supplement your math curriculum and strengthen students' skills.

The two-page activities in *Math Skills Workout Grade 3* are designed to reinforce previously introduced math concepts. Each activity has a colorful teacher page and a skill-based reproducible student page.

The teacher page includes the following:
- the purpose of the activity
- a summary of what students will do
- a list of all needed materials, including any provided patterns
- vocabulary to review before the students complete the reproducible
- two fun-filled extension activities

The student page is a skill-based reproducible that supports NCTM standards. Most reproducibles have a bonus box designed to provide an extra challenge. Answer keys are provided in the back of the book.

Select from a variety of activities to meet your students' individual needs. Then use the accompanying extension activities to provide extra skill reinforcement or to informally assess students' progress. Tailoring math practice has never been so easy!

www.themailbox.com

More great math books from *The Mailbox*®

TEC505.	Lifesaver Lessons® Math • Grade 3
TEC837.	The Best of *The Mailbox*® Math • Grades 1–3
TEC882.	The Best of *Teacher's Helper*® Math • Book 1 • Grades 2–3
TEC3212.	The Best of *Teacher's Helper*® Math • Book 2 • Grades 2–3
TEC1602.	Math Mind Builders • Grade 3

Writers: Amy Barsanti, Heather Godwin, Ann Hefflin, Laura Mihalenko, Valerie Wood Smith, Laura Wagner
Project Manager: Njeri Jones Legrand
Staff Editors: Denine T. Carter, Diane F. McGraw, Deborah G. Swider
Copy Editors: Gina Farago, Karen Brewer Grossman, Karen L. Huffman, Amy Kirtley-Hill, Debbie Shoffner
Cover Artist: Clevell Harris
Art Coordinator: Clevell Harris
Artists: Pam Crane, Theresa Lewis Goode, Nick Greenwood, Clevell Harris, Sheila Krill, Mary Lester, Clint Moore, Kimberly Richard, Greg D. Rieves, Rebecca Saunders, Barry Slate, Donna K. Teal
Typesetters: Lynette Dickerson, Mark Rainey

President, The Mailbox Book Company™: Joseph C. Bucci
Director of Book Planning and Development: Chris Poindexter
Book Development Managers: Stephen Levy, Elizabeth H. Lindsay, Thad McLaurin, Susan Walker
Curriculum Director: Karen P. Shelton
Traffic Manager: Lisa K. Pitts
Librarian: Dorothy C. McKinney
Editorial and Freelance Management: Karen A. Brudnak
Editorial Training: Irving P. Crump
Editorial Assistants: Terrie Head, Melissa B. Montanez, Hope Rodgers, Jan E. Witcher

©2001 THE EDUCATION CENTER, INC.
All rights reserved.
ISBN #1-56234-451-X

Except as provided for herein, no part of this publication may be reproduced or transmitted in any form or by any means, electronic or mechanical, including photocopying, recording, or storing in any information storage and retrieval system or electronic online bulletin board, without prior written permission from The Education Center, Inc. Permission is given to the original purchaser to reproduce patterns and reproducibles for individual classroom use only and not for resale or distribution. Reproduction for an entire school or school system is prohibited. Please direct written inquiries to The Education Center, Inc., P.O. Box 9753, Greensboro, NC 27429-0753. The Education Center®, *The Mailbox*®, *Teacher's Helper*®, the mailbox/post/grass logo, and The Mailbox Book Company™ are trademarks of The Education Center, Inc., and may be the subject of one or more federal trademark registrations. All other brand or product names are trademarks or registered trademarks of their respective companies.

Manufactured in the United States
10 9 8 7 6 5 4 3 2 1

Table of Contents

Number and Operations

Counting Caterpillar Style: Cardinal and ordinal numbers 5–6
Tossing Numbers: Odd and even numbers 7–8
Right on Target: Hundreds, tens, and ones 9–10
Jack's Bean Business: Ones through thousands place 11–12
Are You Ready to Order? Comparing and ordering numbers 13–14
Monstrous Numbers Rally: Numbers through the ten thousands place 15–16
Coast Into Rounding: Rounding numbers 17–18
Home, Sweet Home: Addition and subtraction fact families 19–20
Up and Add 'Em! Two-digit addition with regrouping 21–22
Right on Track! Two-digit addition; three addends 23–24
Busy Bees: Two-digit subtraction with regrouping 25–26
Hefty Subtraction: Three-digit subtraction; regrouping across zero 27–28
A Stitch in "Times": Multiplication 29–30
"Multipli-city": Multiplication 31–32
On a Roll With Division: Division 33–34
Fact Family Airlines: Multiplication and division fact families 35–36
Sailing Along With Fractions: Equal parts of a whole 37–38
Getting a Jump on Fractions: Fractions of a group 39–40
Sweet Comparisons: Comparing fractions 41–42
Putting a Spin on Fractions: Equivalent fractions 43–44
A Menu of Mixed Numbers: Mixed numbers 45–46
Playing a Round of Fractions and Decimals: Fractions and decimals 47–48

Measurement

Snakes of All Sizes: Nonstandard units of length 49–50
Measuring Leaps and Bounds: Customary units of length 51–52
A Royal Ruler: Metric units of length 53–54
A Weighty Decision: Customary units of mass 55–56
Jungle Journey: Metric units of mass 57–58
Fishing for the Right Unit of Capacity: Customary units of capacity 59–60
Ready to Travel? Fahrenheit temperature 61–62
Celsius Sense: Celsius temperature 63–64
A Perimeter Puzzler: Perimeter 65–66
Sammy's Sports Complex: Area 67–68
Under Construction: Finding volume 69–70
Cuckoo for Clocks! Time to the quarter hour 71–72
It's Apple-Pickin' Time! Time to five minutes 73–74
Sweet Times Bakery: Time to the minute 75–76
Time Twists: Elapsed time 77–78
Money in the Bank: Comparing coin sets 79–80
Snack Attack With Money Back: Making change 81–82
Get Set to Own a Pet: Adding and subtracting money 83–84

Geometry

- **Juggling Polygons:** Polygons 85–86
- **Polygon Parade:** Polygons 87–88
- **Shopping for Space Figures:** Space figures 89–90
- **Exploring Space Figures:** Space figures 91–92
- **Dot-to-Dot and Beyond!** Lines, line segments, and rays 93–94
- **Awesome Origami:** Symmetry 95–96
- **Brushing Up on Congruent Figures:** Congruent figures 97–98
- **Angle Antics:** Angles 99–100
- **Geometric Groove:** Slides, flips, and turns 101–102
- **Taxi Takeoff:** Coordinate graphing 103–104
- **Quilt Quest:** Spatial sense 105–106

Probability, Statistics, and Graphing

- **Pep Rally Tally:** Organizing and interpreting data 107–108
- **Survey Celebration:** Organizing and interpreting data 109–110
- **Graph-a-Snack:** Pictographs 111–112
- **Tuning In to Graphing:** Pictographs 113–114
- **Gotta Have Heart!** Bar graphs 115–116
- **Ready, Set, Recycle!** Bar graphs 117–118
- **Splash Into Probability:** Probability 119–120
- **A Probability Picnic:** Probability 121–122
- **Professor Probability:** Probability 123–124
- **What Are the Odds?** Probability 125–126

Algebra

- **Penguin Paintings:** Patterns 127–128
- **Flopsy's Florals:** Patterns 129–130
- **The Case of the Missing Addends:** Missing addends 131–132
- **What's the Scoop?** Missing factors 133–134
- **Cool Calculations:** Commutative property 135–136
- **Arithmetic Roundup:** Using parentheses 137–138
- **Function Junction:** Function tables 139–140

Problem Solving

- **Backward Birdie:** Work backward 141–142
- **Rows of Reasoning:** Logical reasoning 143–144
- **Tricks of the Trade:** Act it out 145–146
- **Flower Power:** Draw a picture 147–148
- **Ready to Race!** Make a table 149–150
- **Score With Patterns!** Find a pattern 151–152
- **Mall Mania:** Make a list 153–154
- **Carnival Capers:** Guess and check 155–156
- **The Race Is On!** Choose the operation 157–158
- **Big Top Strategies:** Choose a strategy 159–160

Patterns 161–168
Answer Keys 169–176

Counting Caterpillar Style

Count on this classy caterpillar to get youngsters' cardinal and ordinal number skills in line!

Purpose: To review cardinal and ordinal numbers

Students will do the following:
- write cardinal and ordinal numbers
- identify numbers as cardinal or ordinal
- use cardinal and ordinal numbers to solve problems

Materials for each student:
- copy of page 6
- pencil

Vocabulary to review:
- cardinal
- ordinal

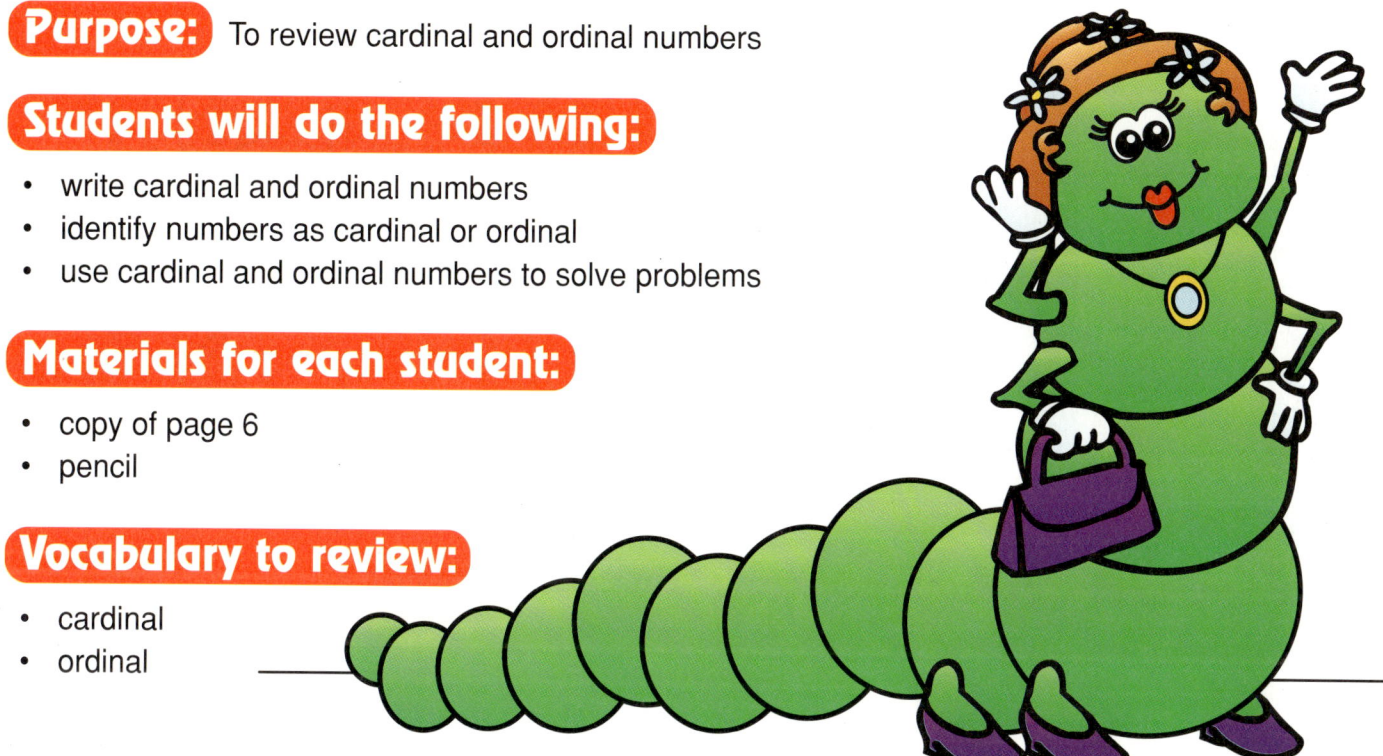

Extension activities to use after the reproducible:

- Take the activity on page 6 a step further to give your youngsters math-related writing practice. Provide each youngster with a sheet of writing paper; then instruct him to study his copy of page 6 and write a detailed description of Carrie Caterpillar. Tell youngsters to include cardinal and ordinal numbers that tell about Carrie's segments and to spell out word names for these numbers as they write. Then have each student trade his completed description with a partner, turn his copy of page 6 facedown, and draw a picture of Carrie Caterpillar on the back based on his partner's description. Next, have partners meet to share their drawings and compare them to the actual picture of Carrie Caterpillar. Students will see which details need to be taken away from or added to their writing at a glance.

- Use this problem-solving learning center to reinforce students' ordinal number skills and ability to extend patterns. Program ten 5" x 7" index cards with challenges similar to the ones shown. Program the back of each card with the correct answer. Laminate the cards for durability. Store the cards, some pencils, and a supply of paper at a center. A student reads the challenge on each card, writes her answer, and then flips the card to check her work.

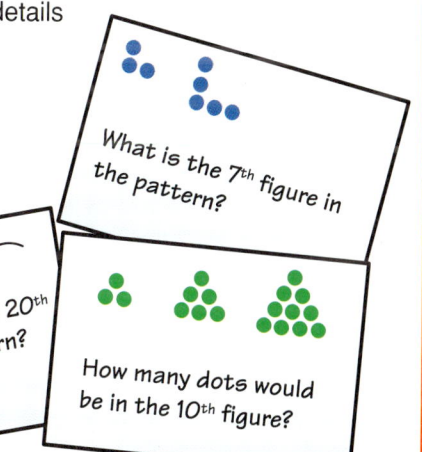

Cardinal and ordinal numbers

Name _____

Cardinal and ordinal numbers

Counting Caterpillar Style

Count Carrie Caterpillar's segments.
Be sure to include Carrie's head!
Use cardinal or ordinal numbers to answer each question.
Then tell if your answers are cardinal or ordinal numbers.
Use the hint box to help you.

Hint Box

Cardinal numbers tell how many, such as 1, 2, or 3.
Ordinal numbers show position in order, such as 1st, 2nd, or 3rd.

1. How many segments are there in all?

2. Which body segments have stars?

3. How many body segments have stripes?

4. Which body segments are not decorated?

5. How many body segments have spots?

6. How many body segments have stars?

7. Which body segments have stripes?

8. How many body segments do not have squares?

9. Which body segment is 4th from the end?

10. How many segments are between the 6th and 13th segments?

11. How many segments are between the 2nd and 12th segments?

12. Which segment is exactly in the middle?

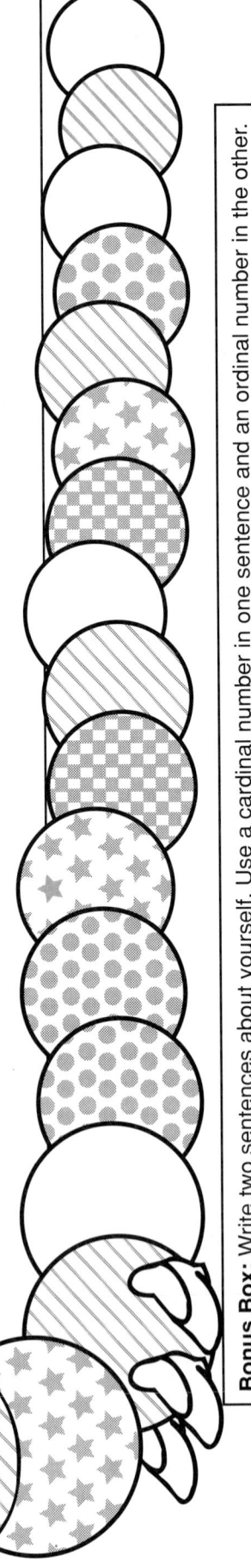

Bonus Box: Write two sentences about yourself. Use a cardinal number in one sentence and an ordinal number in the other.

©2001 The Education Center, Inc. • *Math Skills Workout* • TEC3227 • Key p. 169

Tossing Numbers

Watch students have a ball with odd and even numbers!

Purpose: To identify odd and even numbers

Students will do the following:
- identify numbers as odd or even
- form two- and three-digit odd and even numbers
- add and subtract odd and even numbers

Materials for each student:
- copy of page 8
- pencil
- crayons

Vocabulary to review:
- odd
- even
- sum
- difference

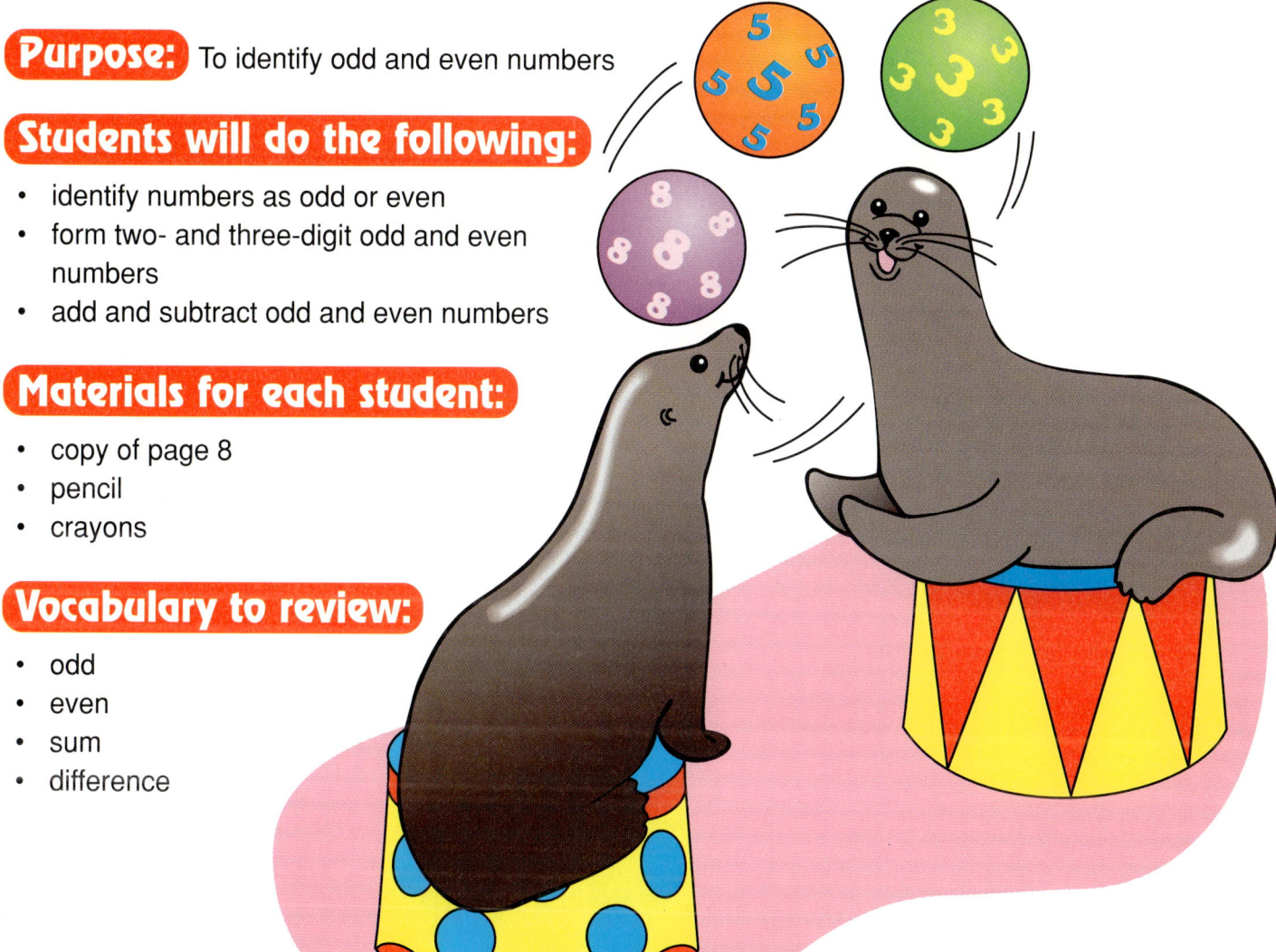

Extension activities to use after the reproducible:
- For more practice forming odd and even numbers, try this cooperative group activity. Provide each student with an index card. Then instruct each child to write a single-digit even number on one side of the card and a single-digit odd number on the other side. Then divide students into groups of three and provide each group with a sheet of paper and a pencil. Give each group a predetermined amount of time to form as many three-digit even numbers as possible by laying their cards faceup on the playing surface. Have students record their numbers on the paper. When time is up, place students into different groups or challenge students in the same group to repeat the process forming odd numbers.

- Here's a partner competition that can't be beat! Pair students and provide each twosome with a copy of the hundreds chart on page 161, a pair of dice, and two different-colored crayons (one for each player). To play the game, each player, in turn, rolls the dice. If the number is even, the player uses his crayon to color any square on the chart with an even number. If the number is odd, the player colors any square on the chart with an odd number. Play continues in this manner. The first player to color five squares in a row—horizontally, vertically, or diagonally—wins the game.

Odd and even numbers

Name _____ Odd and even numbers

Tossing Numbers

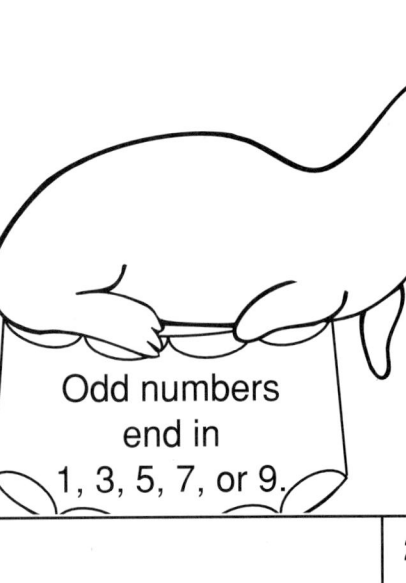

Odd numbers end in 1, 3, 5, 7, or 9.

Check out this amazing act! Use the numbers on the balls to solve the problems below.

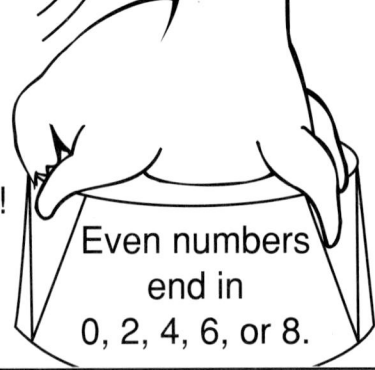

Even numbers end in 0, 2, 4, 6, or 8.

1.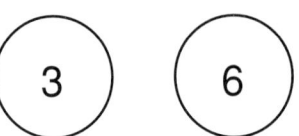

   Form a two-digit odd number. _____

   Form a two-digit even number. _____

2.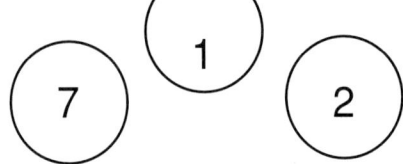

   Form the largest odd three-digit number. _____

3. 0 5 4 8

   Find the sum of the even numbers. _____

4.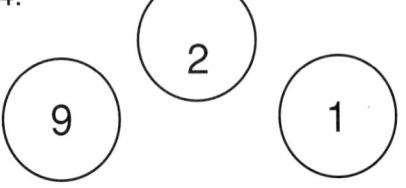

   Form the smallest even three-digit number. _____

5. Answer odd or even.

   8   5

   The sum of these numbers is _____.

   The difference between these numbers is _____.

6. Complete both sentences.

   odd + odd = _____

7. Complete both sentences.

   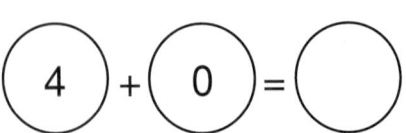

   even + even = _____

8. Complete both sentences.

   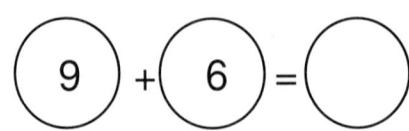

   odd + even = _____

Bonus Box: Color all of the balls with even numbers red. Color all of the balls with odd numbers yellow.

8   ©2001 The Education Center, Inc. • Math Skills Workout • TEC3227 • Key p. 169

Right on Target

Keep your sharpshooters' understanding of hundreds, tens, and ones on target!

Purpose: To interpret place value to the hundreds place

Students will do the following:
- demonstrate an understanding of place value to the hundreds place
- read and write three-digit numbers
- interpret and draw representations of three-digit numbers

Materials for each student:
- copy of page 10
- pencil

Vocabulary to review:
- hundreds
- tens
- ones

Extension activities to use after the reproducible:

- This place-value bingo game hits the mark! Program 30 large index cards as follows: 0 ones–9 ones, 0 tens–9 tens, and 0 hundreds–9 hundreds. Then provide each student with a blank index card and three game markers. Instruct each student to write a three-digit number on his card. Next, shuffle the programmed index cards. Draw and announce the value of each card one at a time. If a student has the corresponding value on his card, he places a game marker on the corresponding number. (For example, if "4 hundreds" is announced, each child who has 4 in the hundreds place covers that number with his game marker.) The first student to cover all three numbers on his card announces "bingo" and becomes the caller for the next round of play.

- Cash in on this behavior plan that reinforces place value! First, explain to students how a dollar, dime, and penny correlate to hundreds, tens, and ones. Then invite youngsters to earn a class reward—such as a popcorn party, no-homework night, or game time—by saving their pennies. Make and laminate a chart similar to the one shown. Then reward good behavior by taping play money in the appropriate sections of the chart. Give students pennies for a good deed or small gesture, dimes for behaviors (such as being prepared for class or quiet transitions), and dollars for whole-group participation (such as the entire class returning its homework). Each time the class earns money, tape the play money coins or dollars in the appropriate columns, regrouping when necessary. Keep a running total until ten dollars is earned. Then present the class with a desired reward.

dollars	dimes	pennies

Hundreds, tens, and ones

Name _____ Hundreds, tens, and ones

Right on Target

Archer Armadillo is a sure shot!
Add up the hundreds, tens, and ones to find the score on each target.

A. Score: _____

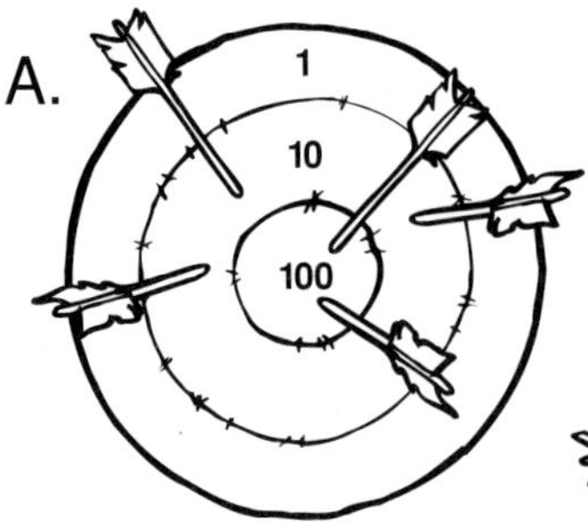

B. Score: _____

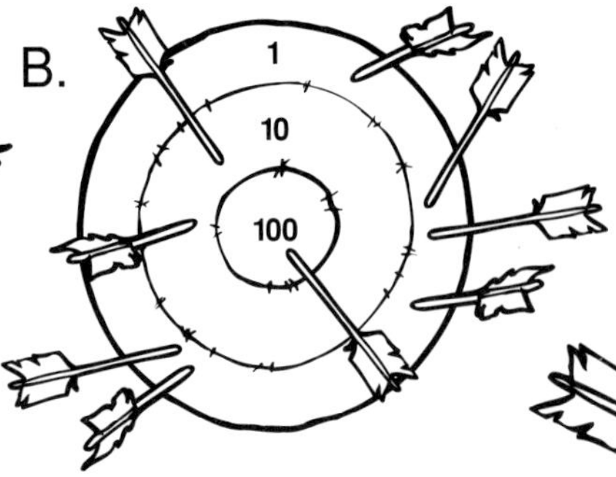

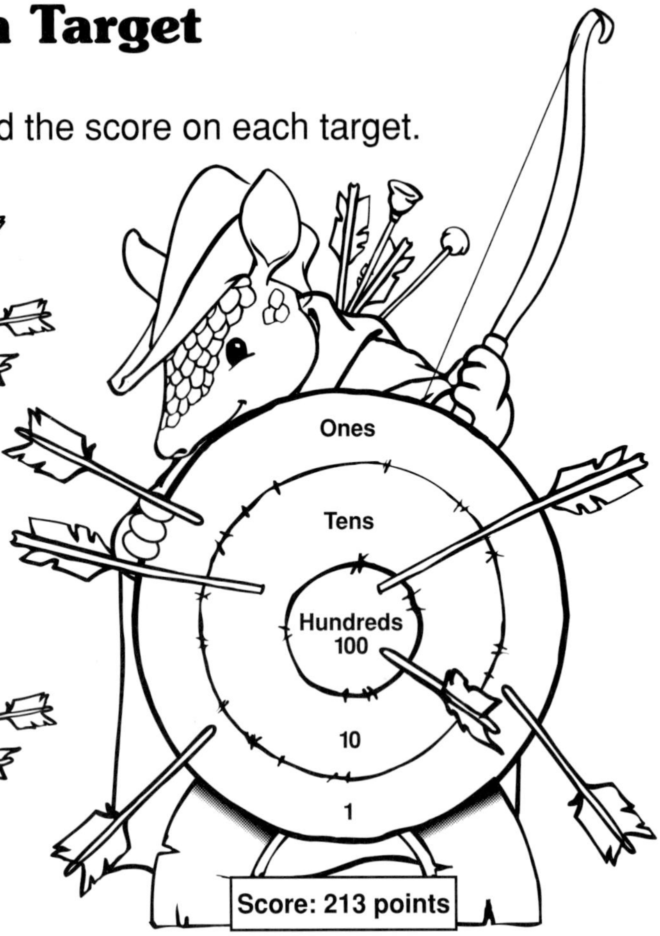

Score: 213 points

C. 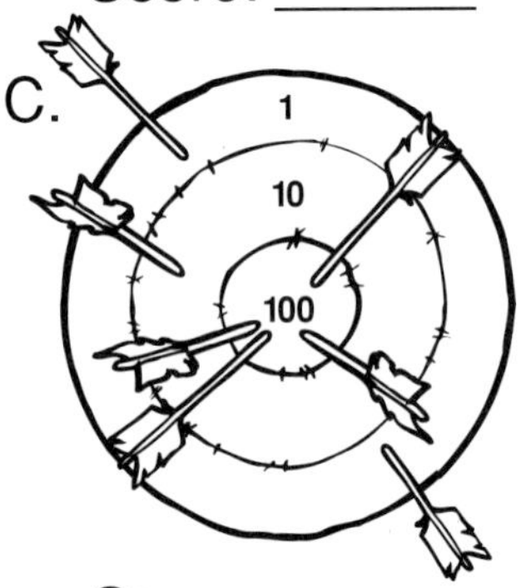 Score: _____

D. Score: _____

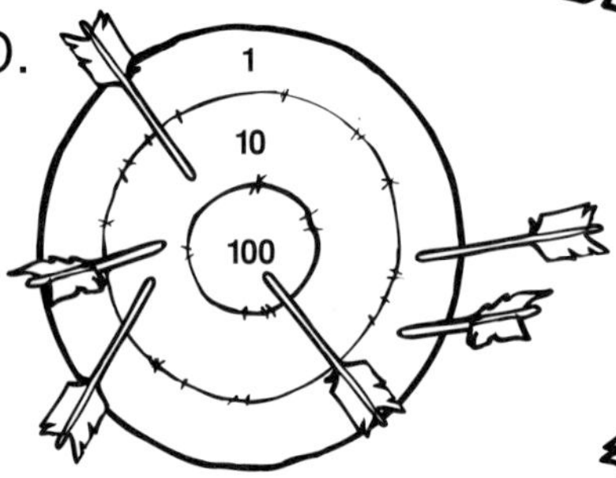

E. Score: _____

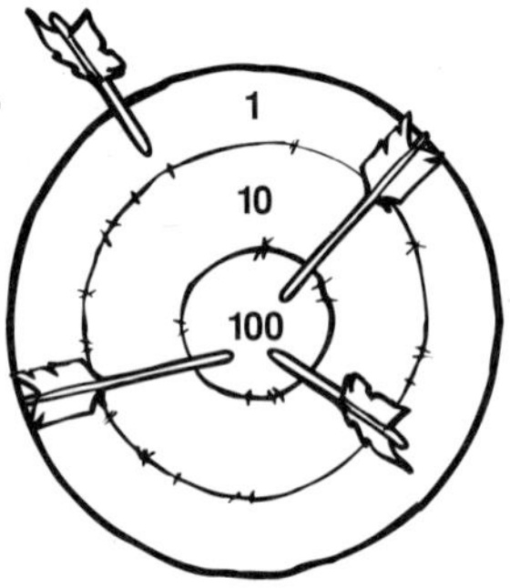

F. Score: _____

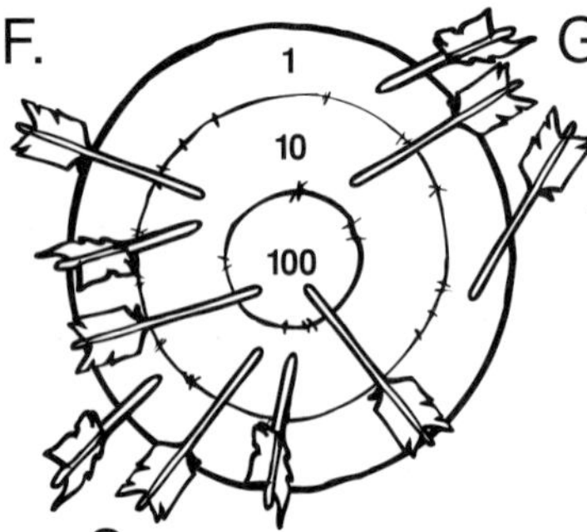

G. Score: _____

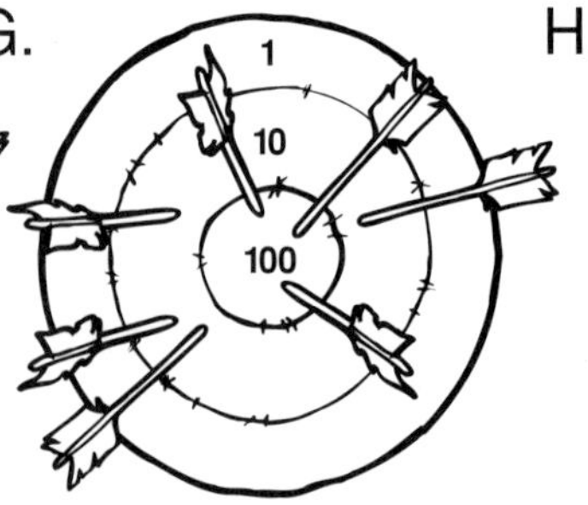

H. Score: _____

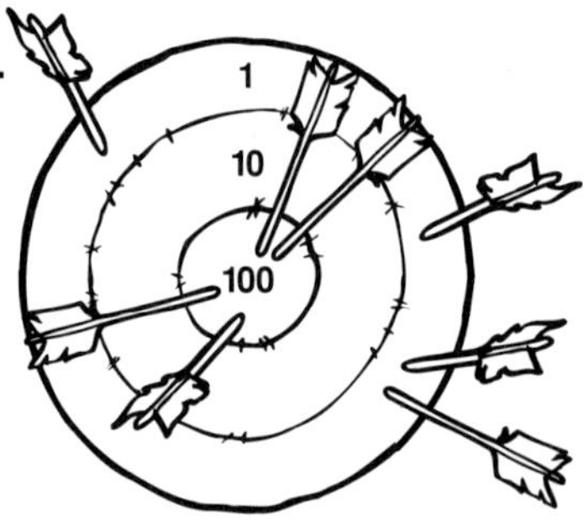

For each target draw arrows to match the score.

I. Score: 314

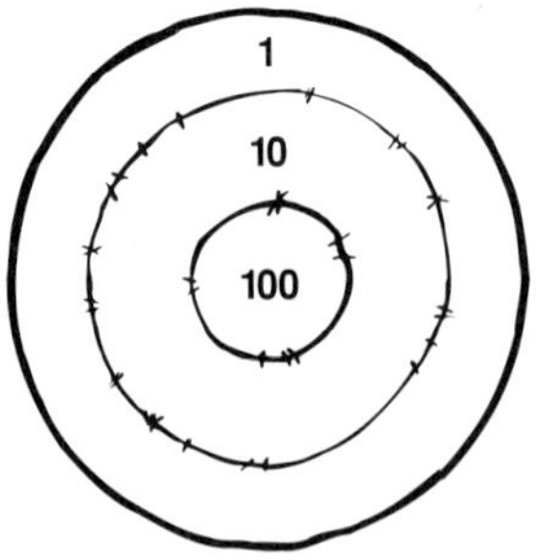

J. Score: 262

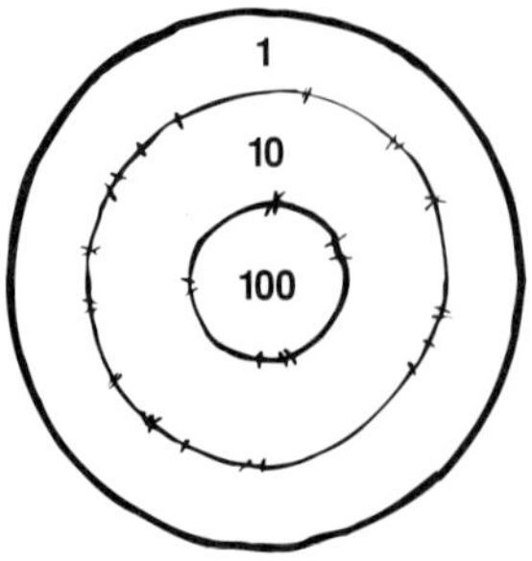

K. Score: 501

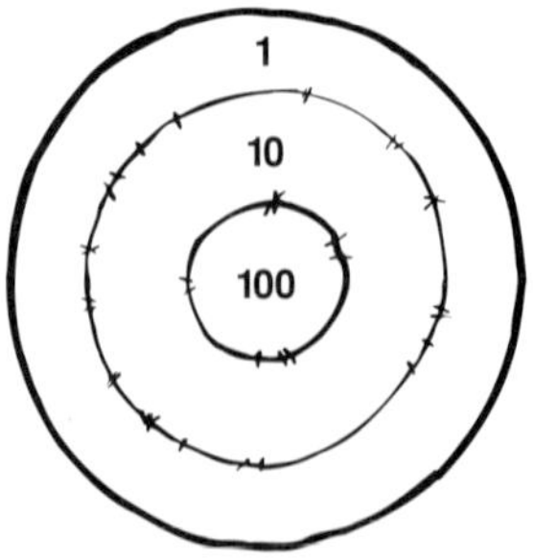

L. Score: 160

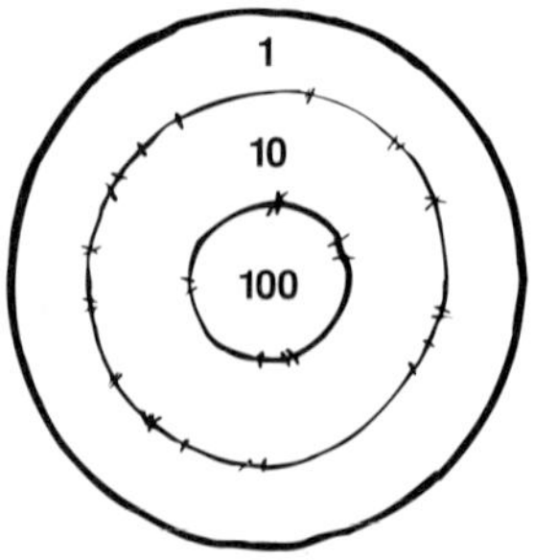

Bonus Box: Above, Archer Armadillo has scored 213 points. If Archer Armadillo wants to score 315 points, where must his next three arrows land?

10   ©2001 The Education Center, Inc. • *Math Skills Workout* • TEC3227 • Key p. 169

Jack's Bean Business

Watch students' understanding of four-digit numbers grow and grow!

Purpose: To determine place value to the thousands place

Students will do the following:
- model numbers to the thousands place
- identify the value of digits in numbers

Materials for each student:
- copy of page 12
- pencil
- scissors

Vocabulary to review:
- place value
- digits

Extension activities to use after the reproducible:
- Clown around with glyphs to give place-value practice an exciting twist! Provide each student with a piece of drawing paper and crayons. Have each student choose any four-digit number and write it at the top of her paper. Then explain to students that a *glyph* is a picture or symbol that conveys information. Tell students that they are going to make glyphs to represent their four-digit numbers. Present students with a code like the one shown. Instruct each student to use the code to illustrate her number. Upon completion, have each student practice writing her number in a variety of different ways, such as writing it in word form, standard form, and expanded form.

- Here's a math exercise your youngsters will love! Announce a four-digit number. Each student writes the number on a piece of paper and then quickly uses the same digits to form and list other four-digit numbers. The student who lists the most possible combinations in the allotted time gets to announce a four-digit number for the next round of play.

Ones through thousands place

Name _____ Ones through thousands place

Jack's Bean Business

Cut out the beans below.
Read the clue under each jar.
Use the beans to form the correct four-digit number.
Write the number on the jar.

Magic Beans for Sale

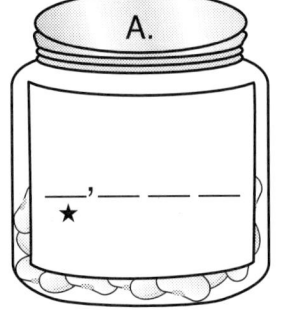

A.
___,___ ___ ___
       ★

The smallest number with 5 in the hundreds place.

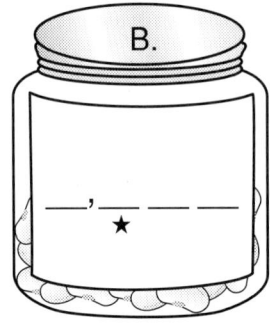

B.
___,___ ___ ___
         ★

The greatest number with 9 in the ones place.

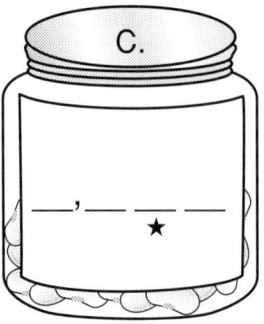

C.
___,___ ___ ___
       ★

The smallest number with 9 in the thousands place.

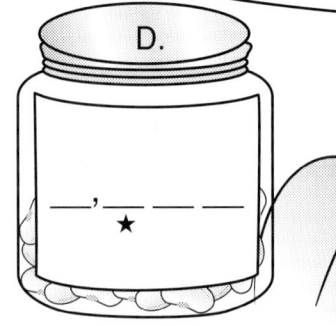

D.
___,___ ___ ___
       ★

The greatest number with 1 in the tens place.

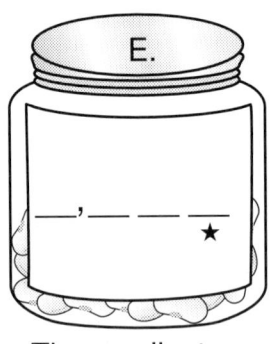

E.
___,___ ___ ___
             ★

The smallest number with 2 in the hundreds place.

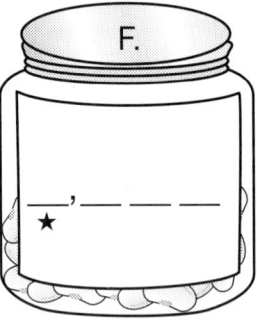

F.
___,___ ___ ___
   ★

The greatest number with 0 in the hundreds place.

G.
___,___ ___ ___
           ★

The smallest number with 4 in the thousands place.

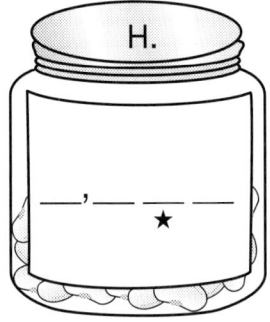

H.
___,___ ___ ___
         ★

The greatest number with 8 in the tens place

Each number above has a ★ under one digit.
Write the value of that digit on the matching line.
(The first one is done for you.)

A. 1,000   B. _____   C. _____   D. _____

E. _____   F. _____   G. _____   H. _____

Bonus Box: What is the largest four-digit number you could make with the beans? What is the smallest?

©2001 The Education Center, Inc. • *Math Skills Workout* • TEC3227 • Key p. 169

0   1   2   3   4   5   6   7   8   9

Are You Ready to Order?

Serve up plenty of numbers. It's ordering time!

Purpose: To order and compare four-digit numbers

Students will do the following:
- compare the value of numbers
- order numbers from least to greatest

Materials for each student:
- copy of page 14
- pencil
- crayons
- scissors
- glue

Vocabulary to review:
- greater than
- less than
- greatest
- least

Extension activities to use after the reproducible:
- Have students compare and order four-digit numbers to create a timeline of events. Find ten important events related to a current topic of study. In random order list each event and the year the event took place. Have students organize the events in chronological order and illustrate a corresponding timeline.

- Have students beat the clock with this fast-paced activity! Divide students into teams of four. Provide each team with a set of number cards (pattern on page 166). Next, call out a four-digit number challenge, such as "Form a number greater than 1,524" or "Form a number between 5,320 and 6,123." Each team discusses and presents a corresponding answer. To present an answer, each team member holds up one of the numerals that makes up the team answer. The first team to answer correctly earns one point. If desired, discuss many possible answers before calling out the next number challenge. The team with the highest score at the end of play wins.

Comparing and ordering numbers

Name _____

Are You Ready to Order?

Comparing and ordering numbers

Help get these orders in order.
Read each order number.
Use the code to color each order.
Cut out the orders and arrange them from *least* to *greatest*.
Glue.

Next order, please!

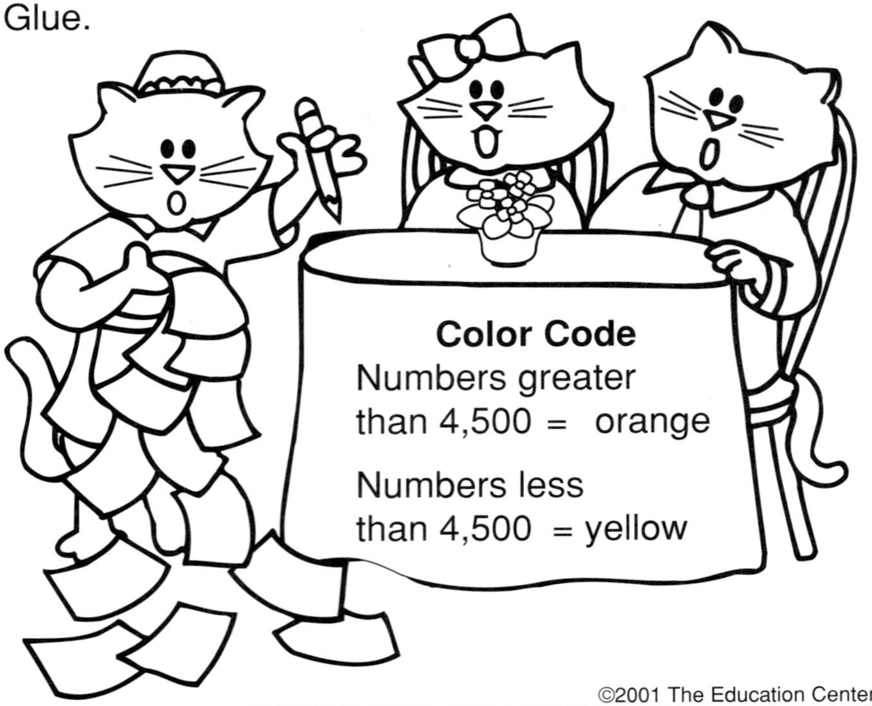

1.	2.	3.	4.	5.	6.
7.	8.	9.	10.	11.	12.
13.	14.	15.	16.	17.	18.

Color Code
Numbers greater than 4,500 = orange
Numbers less than 4,500 = yellow

Bonus Box: Use each symbol (>, <, and =) in a number sentence.

| Order #1,443 | Order #1,972 | Order #8,429 | Order #7,498 | Order #2,673 | Order #4,592 | Order #1,178 | Order #5,000 | Order #7,417 |
| Order #7,280 | Order #1,041 | Order #5,203 | Order #2,628 | Order #3,117 | Order #6,500 | Order #1,999 | Order #3,333 | Order #4,628 |

Monstrous Numbers Rally

Rev up students' ability to read and write monstrous-size numbers!

Purpose: To read and write numbers through the ten thousands place

Students will do the following:
- read number words through the ten thousands place
- write numbers through the ten thousands place

Materials for each student:
- copy of page 16
- pencil

Vocabulary to review:
- ten thousands place

Extension activities to use after the reproducible:
- Get your youngsters on a roll with reading numbers in expanded form and comparing numbers in the thousands! Pair students and provide each pair with a die. Have each student in a pair create a gameboard similar to the one shown. To play a round, a student rolls the die and writes the number shown in any square; then his partner takes a turn. When the four squares on each student's board have been filled, each student writes his final number in standard form on the line. The students then compare numbers. The student with the largest number circles his number. At the end of three rounds, the student with the most numbers circled wins the game!

```
Round 1
☐0,000+☐,000+☐00+☐=_____

Round 2
☐0,000+☐,000+☐00+☐=_____

Round 3
☐0,000+☐,000+☐00+☐=_____
```

- Challenge your students to a game of Guess My Number. In advance, write and conceal a five-digit number. Then challenge your class to guess the number. Have students ask up to 20 questions for which the answer is yes or no. (For example, "Is the number more than 10,000?" or "Is the numeral in the hundreds place nine?") Allow students to record any clues on paper. If students can figure out the number by asking 20 questions or less, the students earn a point. If the students cannot determine the answer in 20 questions, you earn a point. Three points wins the game.

Numbers through the ten thousands place

Name _____

Monstrous Numbers Rally

Read the number below each truck.
Then write the number on the truck.
The first one is done for you.

1. ninety-five thousand, six hundred fifty-seven
2. thirty-one thousand, three hundred twenty-five
3. forty thousand, two hundred ninety-three
4. ten thousand, three hundred
5. fifteen thousand, five hundred fifty-five
6. fifty thousand, two hundred seventeen
7. sixty-four thousand, one hundred seventy-seven
8. twelve thousand, three hundred twenty-eight
9. seventy-nine thousand, one hundred ninety-nine
10. twenty thousand, four hundred twelve

Bonus Box: Use the place value code to color the sections on each truck.

Place Value Code

ones place = yellow
tens place = green
hundreds place = red
thousands place = blue
ten thousands place = orange

Coast Into Rounding

Send students on the ride of their lives as they round numbers to the nearest ten and hundred!

Purpose: To round numbers to the nearest ten or nearest hundred

Students will do the following:
- round numbers to the nearest ten
- round numbers to the nearest hundred

Materials for each student:
- copy of page 18
- pencil
- crayons

Vocabulary to review:
- round
- nearest ten
- nearest hundred

Extension activities to use after the reproducible:
- Use real-life objects to enhance rounding practice. Gather a collection of clean dry-food containers and create a classroom grocery corner. Have students identify the number of grams shown on each container. Then have students round the number to the nearest ten or hundred.

- Step into rounding to the nearest ten (or hundred) with this outdoor activity. Provide each student with a piece of chalk and take youngsters to a paved area. Instruct each youngster to draw two large circles and then write "up" in one circle and "down" in the other. Call out a number to be rounded to the nearest ten (or hundred). Each student decides if the number is rounded up or down and steps into his corresponding circle. Have a student volunteer explain the answer before you call the next number.

Rounding numbers 17

Name _____

Rounding numbers

Coast Into Rounding

Round each number on the chart.
Round two-digit numbers to the nearest ten.
Round three-digit numbers to the nearest hundred.
Find a roller coaster car with the matching answer.
Use the color code on the chart to color each car.

Number	Rounded Number	Color Code
A. 46		red
B. 22		yellow
C. 134		blue
D. 72		green
E. 482		purple
F. 615		orange
G. 13		brown
H. 53		red
I. 156		black
J. 68		green
K. 17		yellow
L. 94		blue
M. 578		orange
N. 237		black
O. 549		purple

Bonus Box: Write three numbers that can be rounded up to 700. Write three numbers that can be rounded down to 700.

Home, Sweet Home

Leap into this "ribbit-ing" addition and subtraction fact-family review!

Purpose: To review addition and subtraction fact families

Students will do the following:
- identify sets of numbers that can be used in fact families
- write fact families
- solve addition and subtraction facts

Materials for each student:
- copy of page 20
- pencil

Vocabulary to review:
- fact family

Extension activities to use after the reproducible:

- For more fact-family practice, provide each student with a 9" x 12" sheet of construction paper, a rubber band, and a pair of scissors. Then have each student follow the directions below to create her very own fact-family challenge.

    Directions:
    1. Fold the construction paper in half four times, unfold it, and then cut the resulting 16 sections apart.
    2. Sort the cards into sets of three (there will be one card left over). For each set, write one digit on each card so that the numbers on all three cards can be used to write a fact family.
    3. Shuffle and stack the number cards.
    4. Title the remaining card "[Name]'s Fact-Family Challenge" and place it on top of the stack. Secure the stack with the rubber band.
    5. Trade cards with a friend. Have your friend sort the cards to find each set of numbers that can make up a fact family.

- Use this partner game to review basic facts. Pair students and provide each pair with a paper bag containing nine tagboard cards numbered 10–18. Then instruct each student to draw and label an addition grid like the one shown. To play the game, each student draws a card in turn, announces the number shown, then returns it to the bag. Each player then finds two addends on the grid (one from each side) that equal the announced sum. He then writes the sum in the corresponding box. The first player to write five sums in a row horizontally, vertically, or diagonally wins the game. For subtraction practice, provide students with a paper bag containing nine tagboard cards numbered 3–11. Have students copy the subtraction grid shown and play the game in the same manner.

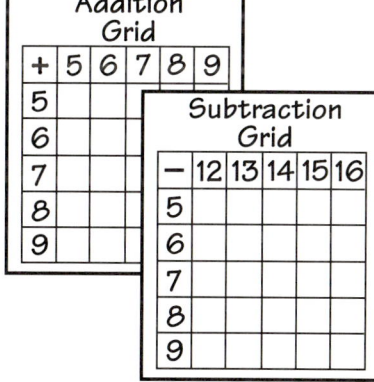

Addition and subtraction fact families

Name _____   Addition and subtraction fact families

Home, Sweet Home

Help these frogs find their lily pad homes.
For each set below, read the numbers on the frogs.
Decide which two numbers can make a fact family
  with the number on each lily pad.
Draw a line from each frog to the matching lily pad.
Then write the fact family. One is done for you.

A. Lily pads: 10, 13. Frogs: 4, 8, 2, 9.

8 + 2 = 10
2 + 8 = 10
10 − 8 = 2
10 − 2 = 8

☐ + ☐ = ☐
☐ + ☐ = ☐
☐ − ☐ = ☐
☐ − ☐ = ☐

B. Lily pads: 12, 15. Frogs: 7, 9, 5, 6.

☐ + ☐ = ☐
☐ + ☐ = ☐
☐ − ☐ = ☐
☐ − ☐ = ☐

☐ + ☐ = ☐
☐ + ☐ = ☐
☐ − ☐ = ☐
☐ − ☐ = ☐

C. Lily pads: 17, 11. Frogs: 5, 8, 9, 6.

☐ + ☐ = ☐
☐ + ☐ = ☐
☐ − ☐ = ☐
☐ − ☐ = ☐

☐ + ☐ = ☐
☐ + ☐ = ☐
☐ − ☐ = ☐
☐ − ☐ = ☐

D. Lily pads: 14, 16. Frogs: 9, 8, 7, 6.

☐ + ☐ = ☐
☐ + ☐ = ☐
☐ − ☐ = ☐
☐ − ☐ = ☐

☐ + ☐ = ☐
☐ + ☐ = ☐
☐ − ☐ = ☐
☐ − ☐ = ☐

Bonus Box: Write your own fact family on the back of this sheet.

©2001 The Education Center, Inc. • Math Skills Workout • TEC3227 • Key p. 169

Up and Add 'Em!

Help students rise to the occasion of successfully adding two-digit numbers with regrouping!

Purpose: To add two-digit numbers with regrouping

Students will do the following:
- add two-digit numbers
- regroup 10 ones as 1 ten
- write two-digit addition equations in columns

Materials for each student:
- copy of page 22
- pencil

Vocabulary to review:
- ones place
- tens place
- regroup

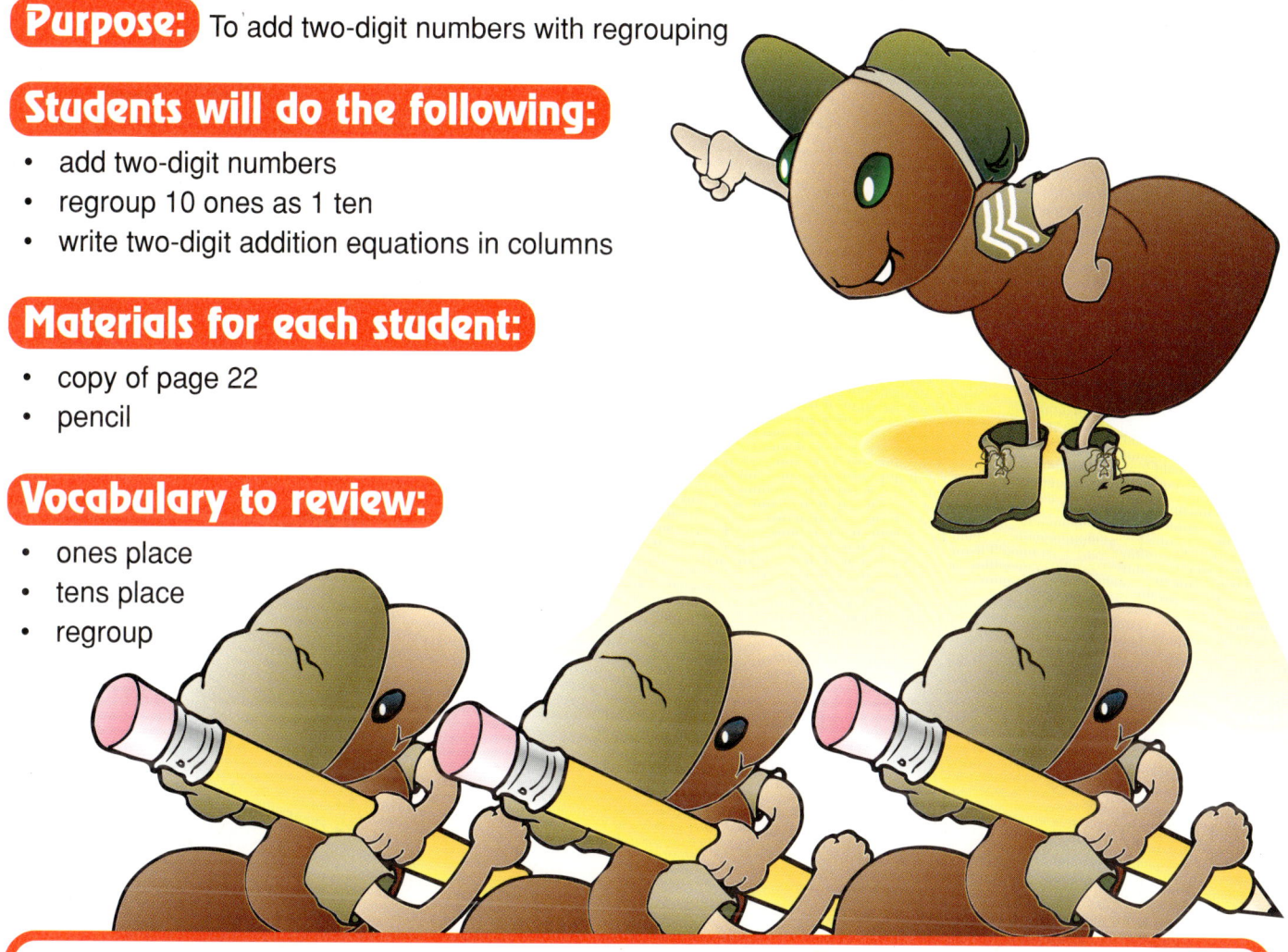

Extension activities to use after the reproducible:

- This tasty activity helps remind students that when they add numbers, the numbers must be regrouped to the next place value. Place students in pairs and give each pair a die, two napkins, and a plate containing a small box of raisins (to represent ones), 20 pretzel sticks (to represent tens), and one square-shaped cookie or cracker (to represent hundreds). Have each partner take turns rolling the die. The student counts out the amount of raisins indicated and places them on his napkin. When he gets ten or more raisins, he must trade them for a pretzel stick. The first student to trade ten pretzel sticks for a cookie or cracker is the winner. After several rounds of play, provide each youngster with a yummy cup of clean raisins and pretzel sticks to eat!

- Turn up the tunes for this melodious addition practice! Have each student write a two-digit number on a large index card. Then, with index card and a crayon in hand, have students march around their classmates' desks while music is playing. When the music stops, each child quickly finds a partner. The pair checks to see if regrouping is necessary to add their numbers. If so, each student tallies two points on the back of her index card. If not, or if a student is unable to pair with a partner, each student gets one point. Continue play in this manner with each student seeking a different partner each time.

Two-digit addition with regrouping

Name _____   Two-digit addition with regrouping

Up and Add 'Em!

Don't forget to regroup, troops!

Write each problem.
Add the numbers.
The first one is done for you.

A. 15 + 39 +
$$\begin{array}{r}15\\+\ 39\\\hline 54\end{array}$$

B. 17 + 68 + _____

C. 48 + 25 + _____

D. 21 + 39 + _____

E. 26 + 45 + _____

F. 48 + 28 + _____

G. 15 + 36 + _____

H. 27 + 56 + _____

I. 17 + 29 + _____

J. 34 + 38 + _____

K. 47 + 28 + _____

L. 27 + 37 + _____

Bonus Box: Write the sums in order from greatest to least.

Right on Track!

Set youngsters' addition skills in motion. All aboard!

Purpose: To reinforce two-digit addition with more than two addends

Students will do the following:
- add three two-digit numbers
- regroup when adding

Materials for each student:
- copy of page 24
- pencil

Vocabulary to review:
- addend
- regroup
- sum

Extension activities to use after the reproducible:

- Reinforce students' addition skills with this sweet activity! Provide each student with a flat cellophane-wrapped lollipop and a fine-point permanent marker. Instruct each student to write a two-digit addition problem with three addends on his lollipop. Then have him write the answer to his problem on the opposite side. Ask each student to trade lollipops with a partner, solve his partner's problem, and then flip the lollipop to check his answer. Have students repeat this process with new partners until each student has solved ten problems. Upon completion, allow students to savor their lollipop treats.

- This addition practice is sure to put students' problem-solving talents to work! Write a problem on the chalkboard similar to the one shown. Divide students into groups of three. Challenge each threesome to find ten possible ways to solve the problem using different numbers in the boxes each time.

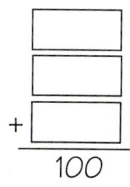

Two-digit addition, three addends

Name _____ Two-digit addition, three addends

Right on Track!

Solve each problem.
Cross out each sum in the answer bank as you use it.

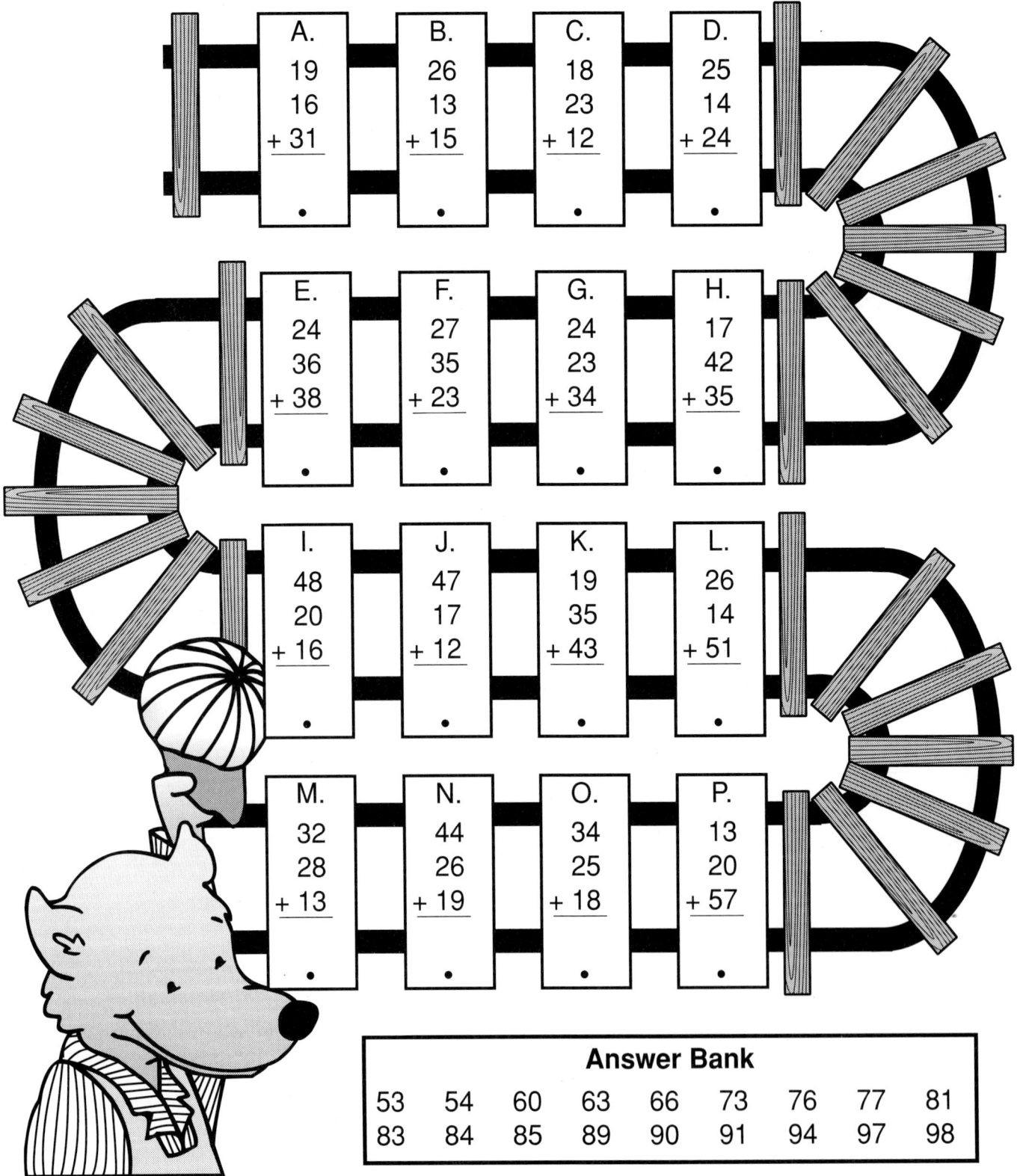

A. 19 16 + 31	B. 26 13 + 15	C. 18 23 + 12	D. 25 14 + 24
E. 24 36 + 38	F. 27 35 + 23	G. 24 23 + 34	H. 17 42 + 35
I. 48 20 + 16	J. 47 17 + 12	K. 19 35 + 43	L. 26 14 + 51
M. 32 28 + 13	N. 44 26 + 19	O. 34 25 + 18	P. 13 20 + 57

Answer Bank

| 53 | 54 | 60 | 63 | 66 | 73 | 76 | 77 | 81 |
| 83 | 84 | 85 | 89 | 90 | 91 | 94 | 97 | 98 |

Bonus Box: Circle the two numbers in the answer bank that were not used. Then write three addends that total each number.

Busy Bees

Get your busy little bees all abuzz about subtraction!

Purpose: To subtract two-digit numbers with regrouping

Students will do the following:
- subtract two-digit numbers
- use place value to regroup

Materials for each student:
- copy of page 26
- pencil
- crayons
- scissors
- glue

Vocabulary to review:
- ones place
- tens place
- regroup

Extension activities to use after the reproducible:
- Bring out the dominoes for a fun subtraction game. Pair students and give each pair a set of dominoes (or a construction paper copy of the dominoes on page 163). Instruct the players to place their dominoes facedown on the playing surface. To play, Player 1 chooses two dominoes and turns each domino horizontally to read a two-digit number. Then she uses the resulting numbers to write and solve a subtraction problem. If she has to regroup to solve the problem, she gets to keep the dominoes. If she does not have to regroup, she returns the dominoes to the playing surface. Player 2 takes a turn in the same manner. Alternate play continues until one player earns ten dominoes to win the game.

- This "class-y" writing activity will inspire your youngsters to share their subtraction knowledge! Ask students to brainstorm what they have learned about subtraction with regrouping. Then ask students to think of advice they would give a younger student who is just learning how to subtract and regroup. Instruct students to write a letter that begins "Dear Student," to explain the process. Remind students to include examples in their letters. Compile students' helpful letters into a class book to share with your youngsters next year!

Two-digit subtraction with regrouping

Name_____ Two-digit subtraction with regrouping

Busy Bees

Solve each problem. Show your work.
If the answer on the hive is correct, color the hive.
If the answer on the hive is incorrect, copy and answer the problem on a blank hive.
Color and cut out that hive and glue it on top of the incorrect one.

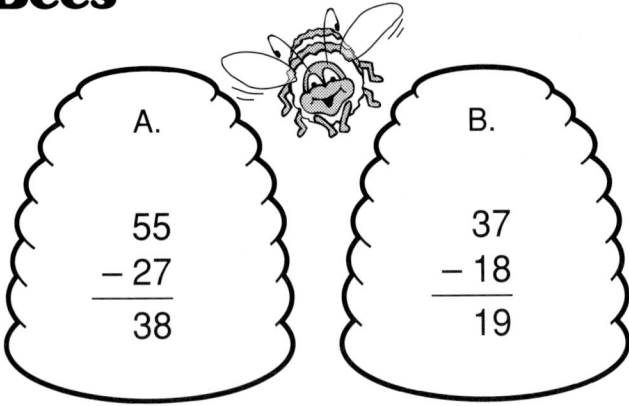

A.
$$\begin{array}{r} 55 \\ -27 \\ \hline 38 \end{array}$$

B.
$$\begin{array}{r} 37 \\ -18 \\ \hline 19 \end{array}$$

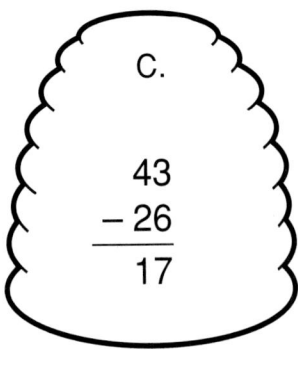

C.
$$\begin{array}{r} 43 \\ -26 \\ \hline 17 \end{array}$$

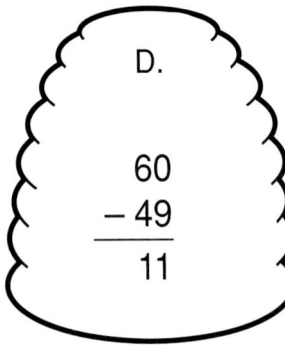

D.
$$\begin{array}{r} 60 \\ -49 \\ \hline 11 \end{array}$$

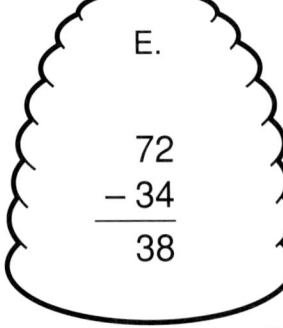

E.
$$\begin{array}{r} 72 \\ -34 \\ \hline 38 \end{array}$$

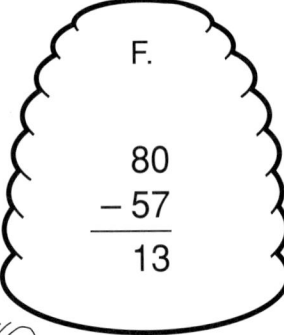

F.
$$\begin{array}{r} 80 \\ -57 \\ \hline 13 \end{array}$$

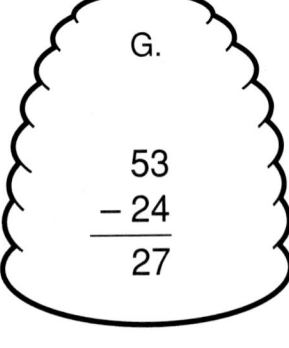

G.
$$\begin{array}{r} 53 \\ -24 \\ \hline 27 \end{array}$$

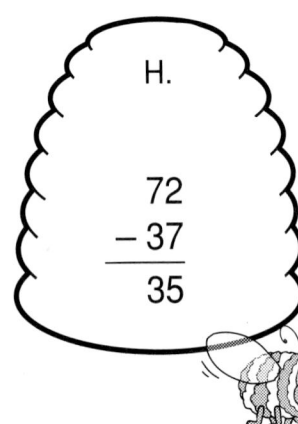

H.
$$\begin{array}{r} 72 \\ -37 \\ \hline 35 \end{array}$$

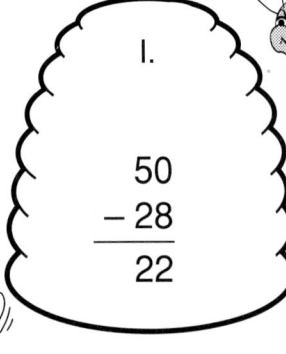

I.
$$\begin{array}{r} 50 \\ -28 \\ \hline 22 \end{array}$$

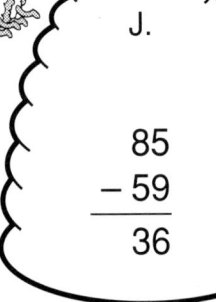

J.
$$\begin{array}{r} 85 \\ -59 \\ \hline 36 \end{array}$$

Bonus Box: Write a word problem that involves subtracting and regrouping two-digit numbers.

©2001 The Education Center, Inc. • Math Skills Workout • TEC3227 • Key p. 170

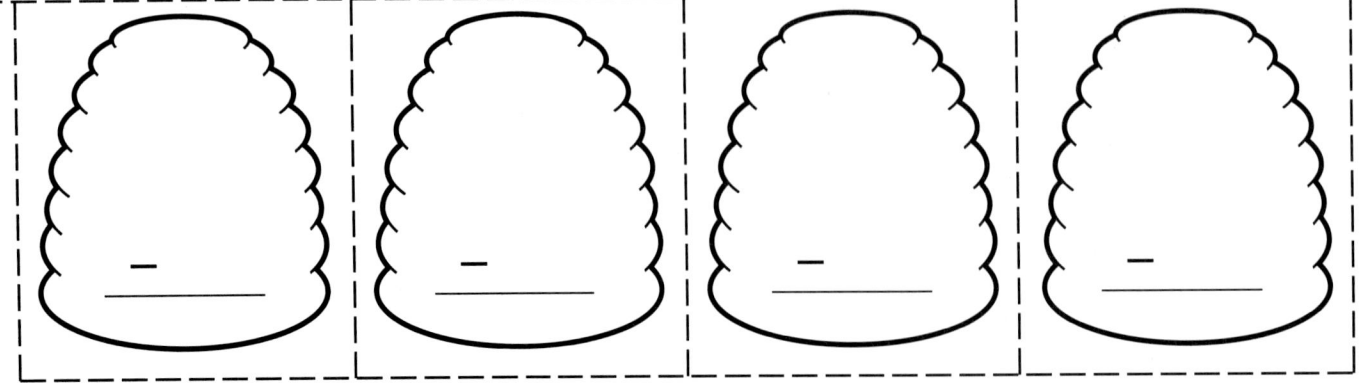

Hefty Subtraction

This powerful subtraction practice is sure to lift students' success to a record high!

Purpose: To solve three-digit subtraction problems with regrouping across zero

Students will do the following:
- subtract three-digit numbers
- regroup tens and ones
- use place value to subtract with zeros

Materials for each student:
- copy of page 28
- pencil

Vocabulary to review:
- place value
- regrouping

Extension activities to use after the reproducible:

- Try this variation of the popular game tic-tac-toe. Arrange nine chairs to resemble a tic-tac-toe grid. Make a pair of number cubes (use pattern on page 162) with three-digit numbers containing zero on each side and a set of ten cards—half programmed with X and half with O. Divide students into two teams, X and O; then roll the dice. Using the numbers rolled, each student writes and solves the subtraction problem. Choose a member of Team X to state the answer. If he answers correctly, he gets a card displaying his team's name and sits in the chair of his choice. If he answers incorrectly, a member of Team O gets a chance to state the answer and choose a chair. Alternate play continues until one team gets three students in a row and wins the round. Play additional rounds, making sure that each student has a chance to answer for his team.

- For more subtraction practice, present students with a table like the one shown. Have students study the table to answer a variety of questions that encourage subtraction, such as "How much farther is Atlanta, Georgia, than New York, New York, from Washington, DC?" and "What is the difference between the distance to St. Louis, Missouri, and Pittsburgh, Pennsylvania, from Washington, DC?" Then have students write and solve their own word problems based on the table.

Approximate Mileage From Washington, DC	
City	Distance
Atlanta, GA	630
Memphis, TN	908
New York, NY	225
Pittsburgh, PA	230
St. Louis, MO	801

Three-digit subtraction, regrouping across zero

Hefty Subtraction

Three-digit subtraction, regrouping across zero

Which day of the week is the best day to enter a weight-lifting competition?
To find out, solve each problem.
Then match each letter to a numbered line below.

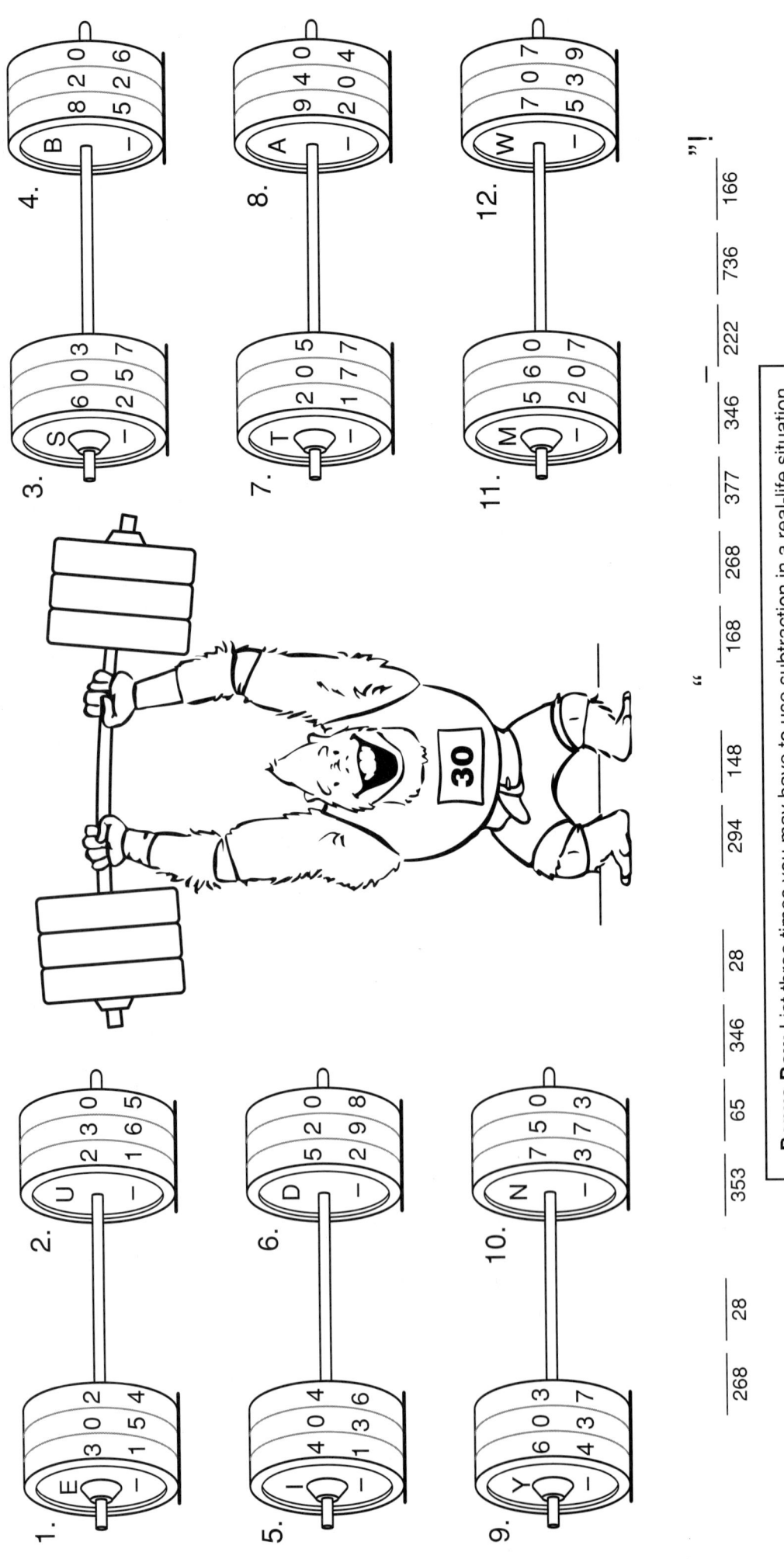

1. E 302 − 154
2. U 230 − 165
3. S 603 − 257
4. B 820 − 526
5. — 404 − 136
6. D 520 − 298
7. T 205 − 177
8. A 940 − 204
9. Y 603 − 437
10. N 750 − 373
11. M 560 − 207
12. W 707 − 539

268 28   353 65   346 28   294 148 168 268 377 346 222 736 166 "!"

Bonus Box: List three times you may have to use subtraction in a real-life situation.

A Stitch in "Times"

Make learning multiplication "sew" much fun!

Purpose: To reinforce the concept of multiplication

Students will do the following:
- count sets and members of sets
- write multiplication sentences

Materials for each student:
- copy of page 30
- pencil

Vocabulary to review:
- multiplication

Extension activities to use after the reproducible:

- Here's a fact review game that's perfect for small groups. Give each group 36 index cards. Instruct students to number each set of nine cards from 2 to 10. One student collects and shuffles the cards, and then deals two cards to each player. Each player multiplies the numbers shown on his cards. The player with the highest product wins the round. The winner collects, shuffles, and deals the cards for the next round of play.

- Create this one-of-a-kind quilt to give your youngsters a patchwork of multiplication practice! Provide each student with a copy of the centimeter graph paper on page 167. List up to eight multiplication facts on the chalkboard. Instruct each student to outline an array on her graph paper for each problem and to write the multiplication sentence inside the array. Upon completion, have each student color and cut out her arrays. Then have her arrange her arrays to form a desired picture or pattern and glue them on a six-inch square of white construction paper. Next, have each child glue her completed square in the center of an eight-inch square of colorful construction paper that you have punched with a series of equally spaced holes—four per side. To assemble the quilt, have student volunteers loop several one-foot lengths of yarn through the holes to stitch the resulting projects into a desired quilt shape. Provide assistance as needed.

Multiplication

Name _____   Multiplication

A Stitch in "Times"

Miss Gertrude is stitching designs to represent her favorite months of the year.
Study the cross-stitches in each design to complete the sentences.
(Hint: Each **X** equals one cross-stitch.)
Then write a matching multiplication sentence.

1.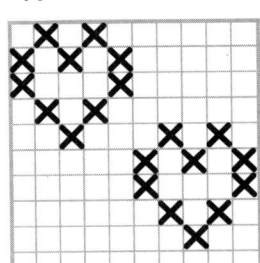
   There are ___ hearts.
   Each heart has ___ cross-stitches.

   ___ x ___ = ___

2.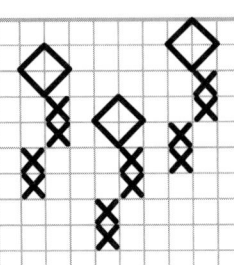
   There are ___ kite tails.
   Each kite tail has ___ cross-stitches.

   ___ x ___ = ___

3.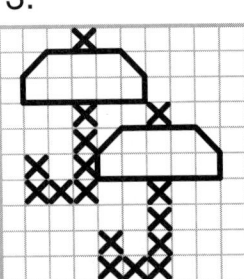
   There are ___ umbrellas.
   Each umbrella has ___ cross-stitches.

   ___ x ___ = ___

4.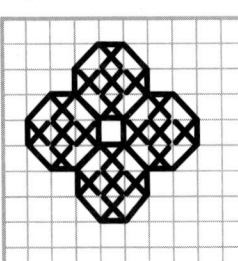
   There are ___ flower petals.
   Each petal has ___ cross-stitches.

   ___ x ___ = ___

5.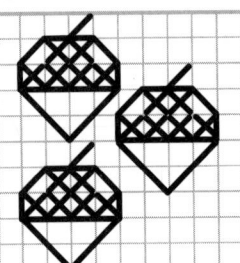
   There are ___ acorn tops.
   Each acorn top has ___ cross-stitches.

   ___ x ___ = ___

6.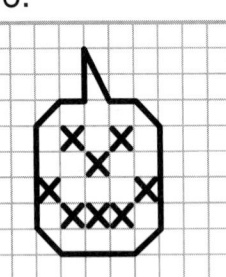
   There is ___ jack-o'-lantern.
   Each jack-o'-lantern has ___ cross-stitches.

   ___ x ___ = ___

7.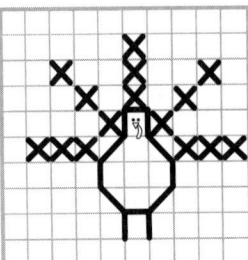
   There are ___ turkey feathers.
   Each feather has ___ cross-stitches.

   ___ x ___ = ___

8.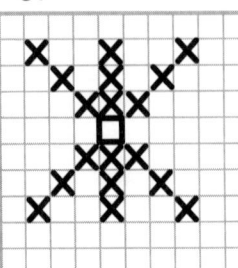
   There are ___ points on the snowflake.
   Each point has ___ cross-stitches.

   ___ x ___ = ___

Bonus Box: Which months are Miss Gertrude's favorites? Write the month that each design above represents.

"Multipli-city"

Use this array of ideas to reinforce students' understanding of multiplication!

Purpose: To reinforce the concept of multiplication

Students will do the following:
- write multiplication facts shown by arrays
- use factors to draw arrays

Materials for each student:
- copy of page 32
- pencil

Vocabulary to review:
- array
- factor
- product

Extension activities to use after the reproducible:

- Here's a fun way for students to improve rapid and accurate recall of multiplication facts. Have groups of four or five students form a circle. Provide each group with a clipboard displaying a multiplication fact drill sheet and a pencil. At the designated start, the first student in each group answers the first multiplication fact and quickly passes the clipboard and pencil to the next student. The next student answers the next fact or erases the previous answer if it is incorrect and passes the clipboard. Play continues in this manner for a predetermined amount of time. Provide each group with a corresponding answer key to check its work and record the number of correct problems. To play the game again, provide each group with a different fact sheet. Challenge each group to improve its previous score.

- Lead your youngsters to investigate patterns in multiples. Make a transparency of the hundreds chart on page 161. Display the transparency, give a student volunteer five beans or other markers, and ask him to place them on the first five multiples of the number 2 (lead the student through the 2 times table, if necessary, to help him identify the multiples). Tell students that a visual pattern will begin to appear on the hundreds chart. Have each student predict what the pattern would be if all of the multiples of 2 were covered and then write his prediction. Have a new student volunteer cover the remaining multiples of 2 to reveal the pattern. Ask each student to record the actual outcome. Repeat this process to explore patterns of different multiples.

Name _____   Multiplication

"Multipli-city"

How many windows must Pete Pigeon wash?
Look at the array of windows on each building.
On the car below each building, write the
multiplication fact the array shows.
Pete Pigeon has done the first one for you.

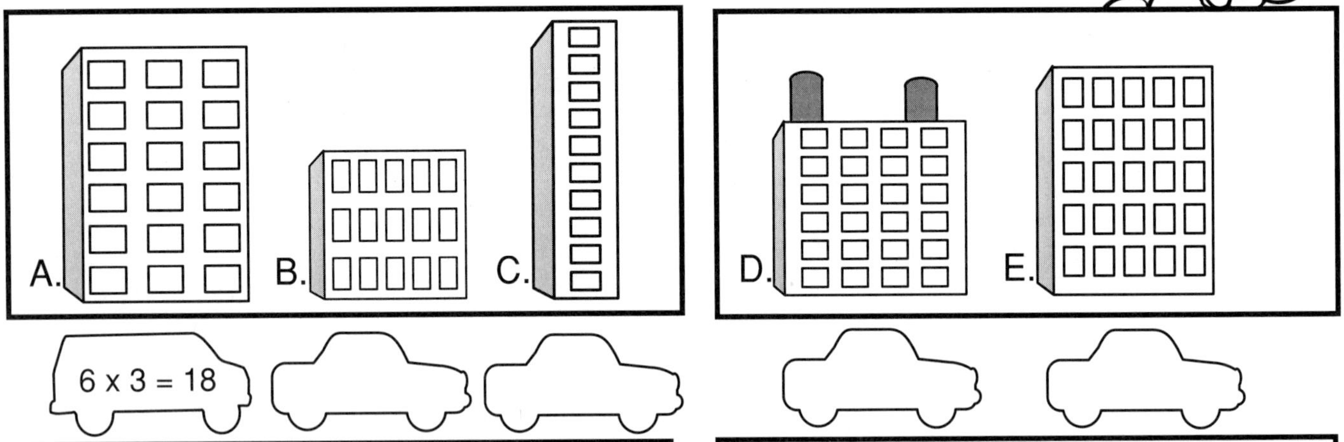

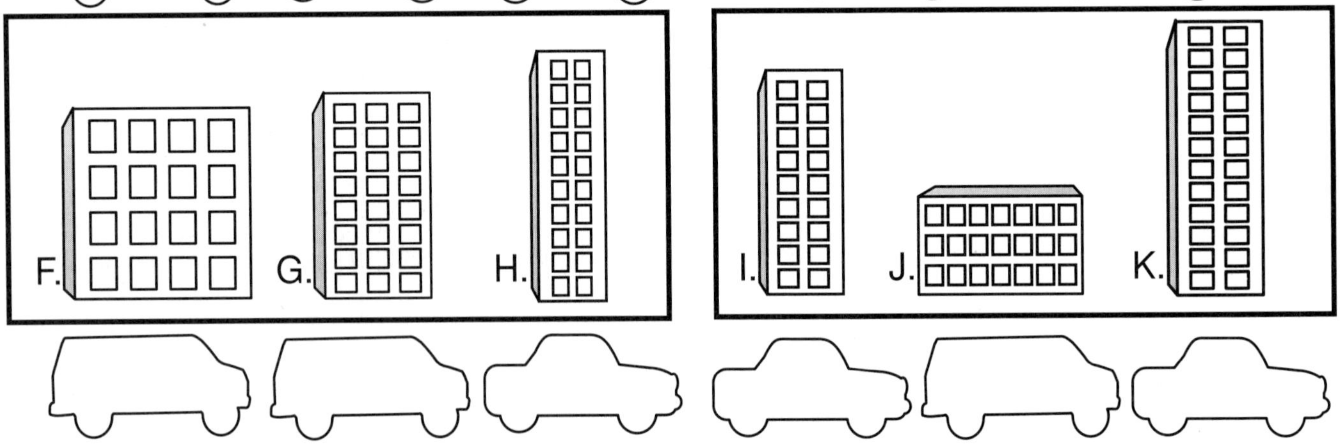

The array of windows has 6 rows and 3 columns.

A. B. C.   6 × 3 = 18

D. E.

F. G. H.

I. J. K.

Now look at the multiplication fact on each car. Draw a matching array of windows on the building and complete the matching fact.

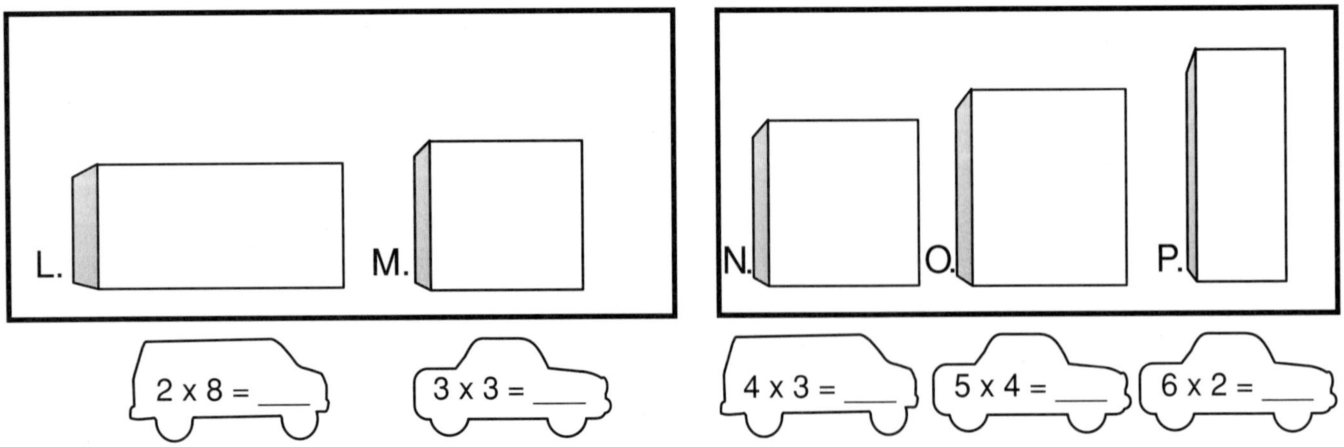

L. M.   2 × 8 = ___   3 × 3 = ___

N. O. P.   4 × 3 = ___   5 × 4 = ___   6 × 2 = ___

Bonus Box: Draw a building with an array of 14 windows. Write a matching multiplication fact.

On a Roll With Division

Use this "marble-ous" activity to get your youngsters on a roll with division!

Purpose: To reinforce the concept of division

Students will do the following:
- divide manipulatives into equal groups
- write division sentences

Materials for each student:
- copy of page 34
- scissors
- pencil

Vocabulary to review:
- divide
- dividend
- divisor
- quotient

Extension activities to use after the reproducible:

- Use these challenging problems to give your youngsters a leg up on division! Write the math problems below on the chalkboard. Ask each student to write a corresponding division sentence. Encourage youngsters to draw illustrations to help them solve each problem. Upon completion, have each student write and solve similar math problems of his own.

   12 legs—how many ants? *(12 ÷ 6 = 2)*
   16 legs—how many dogs? *(16 ÷ 4 = 4)*
   6 legs—how many birds? *(6 ÷ 2 = 3)*
   24 legs—how many spiders? *(24 ÷ 8 = 3)*

- Students can sink their teeth into this manipulative activity. Give each student 12 plastic chips that represent a dozen tasty doughnuts. Have each student manipulate her doughnuts to figure out how they can be shared equally between two people and write a corresponding division sentence. Then have students repeat this process to find out how the doughnuts can be shared equally between 1, 3, 4, 6, and 12 people. Now that youngsters understand how to equally share doughnuts, reward them with a delicious doughnut snack!

Division

Name _____   Division

On a Roll With Division

Take the marble challenge!
Cut out the marbles.
Use them to help you complete the table.
The first row is completed for you.

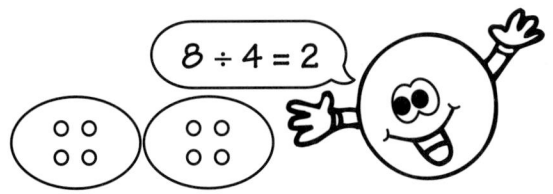

Start With…	Number to Put in Each Group	Number of Groups You Get	Matching Illustration	Matching Division Problem
8 marbles	2	4	○○ ○○ ○○ ○○	8 ÷ 2 = 4
9 marbles	3			
10 marbles	2			
15 marbles	5			
12 marbles	6			
16 marbles	4			
14 marbles	7			
12 marbles	4			
16 marbles	8			
18 marbles	6			

Bonus Box: Draw three different ways that 18 marbles can be divided into equal groups.

©2001 The Education Center, Inc. • *Math Skills Workout* • TEC3227 • Key p. 170

Fact Family Airlines

Get on board this nonstop flight to multiplication and division success!

Purpose: To explore multiplication and division fact families

Students will do the following:
- identify sets of numbers that can be used in fact families
- write fact families
- solve multiplication and division facts

Materials for each student:
- copy of page 36
- pencil

Vocabulary to review:
- fact family

Extension activities to use after the reproducible:

- Challenge your super sleuths to solve a case of missing numbers! Using chalk, draw the outlines of several magnifying glass shapes. Inside each magnifying glass, write a fact family that reveals the same number in each equation as shown. Challenge students to find several answers for each magnifying glass in a given amount of time. When time is up, call students to the board to share possible answers.

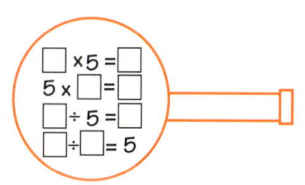

- Provide each student with a large index card, markers, and scissors. Have each student follow the directions below to make a fact family puzzle. Then store completed puzzle pieces in a decorated envelope at a center for students to assemble as they review fact families.

    ### Directions:
    1. Brainstorm a set of numbers that make a fact family. Write the numbers on the front of the card and decorate the card.
    2. On the back of the card, write a corresponding fact family.
    3. Cut apart the numbers in puzzle-piece fashion, separating the individual numbers in the set.

Multiplication and division fact families 35

Name _____ Multiplication and division fact families

Fact Family Airlines

Read each set of numbers.
If the set can be used to make a fact family, write the fact family on the suitcase with the matching letter.
If the set cannot be used to make a fact family, circle the set.

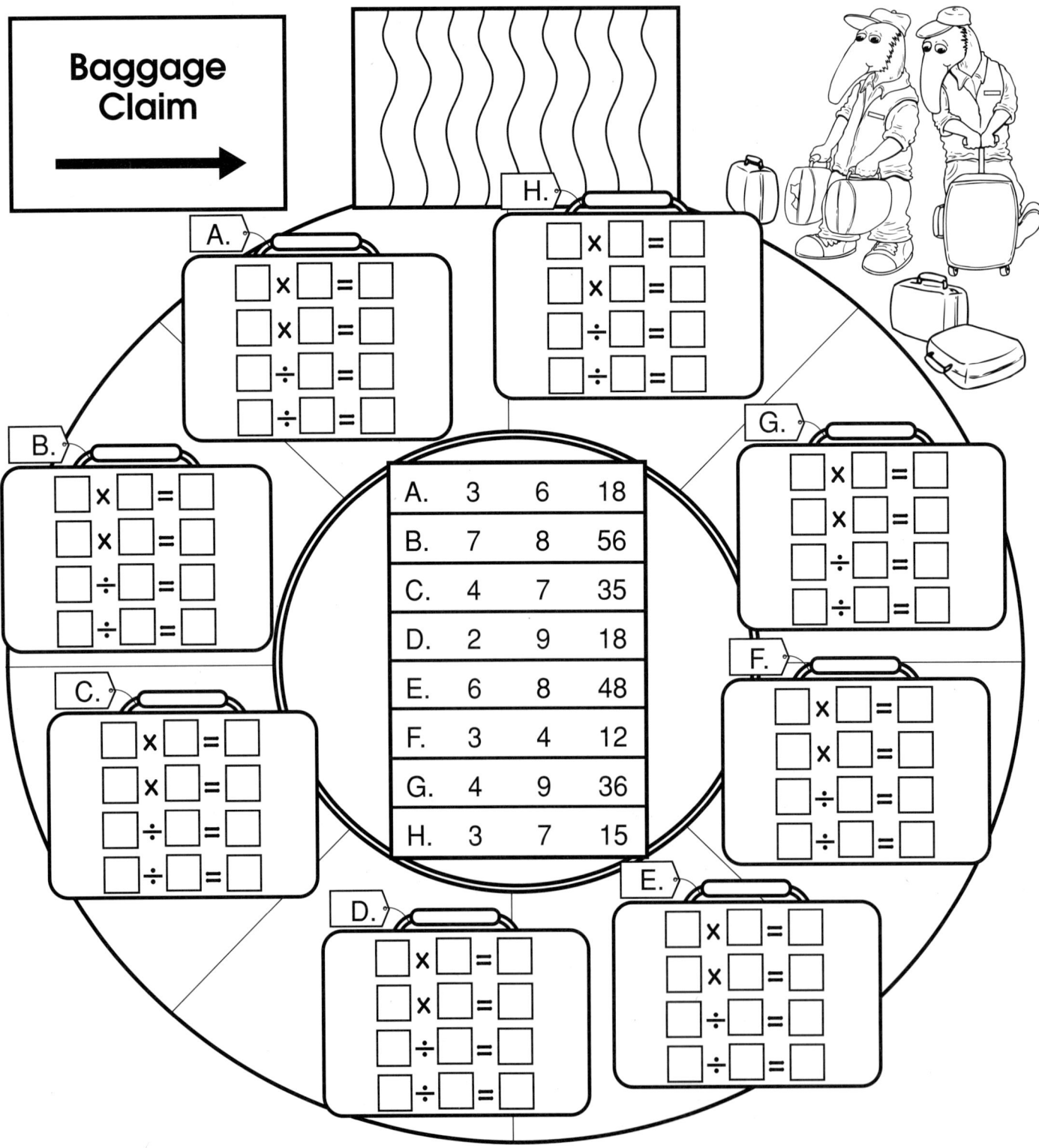

A.	3	6	18
B.	7	8	56
C.	4	7	35
D.	2	9	18
E.	6	8	48
F.	3	4	12
G.	4	9	36
H.	3	7	15

Bonus Box: For each circled set of numbers, change one number so that the set can make a fact family. Then write the fact family on the matching suitcase.

Sailing Along With Fractions

Sail into your study of fractions by having students identify equal parts of a whole!

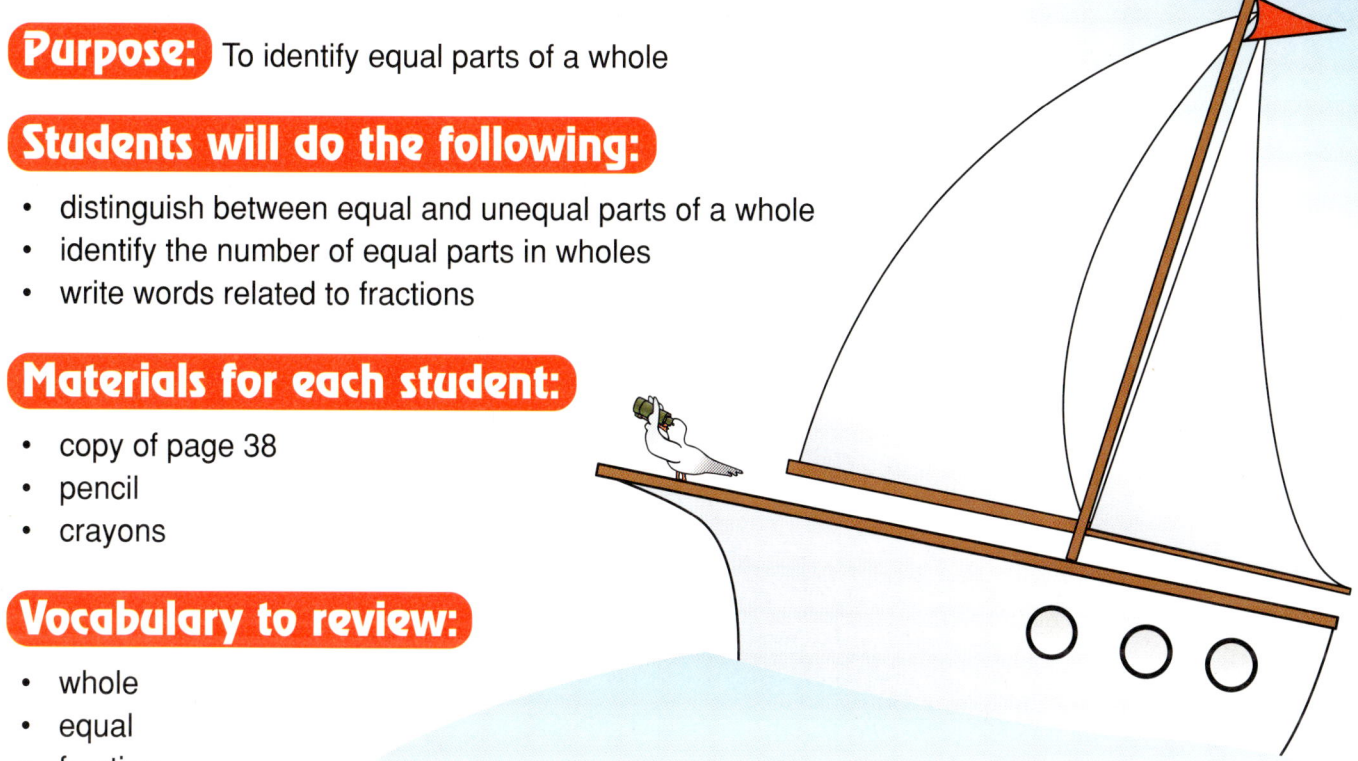

Purpose: To identify equal parts of a whole

Students will do the following:
- distinguish between equal and unequal parts of a whole
- identify the number of equal parts in wholes
- write words related to fractions

Materials for each student:
- copy of page 38
- pencil
- crayons

Vocabulary to review:
- whole
- equal
- fraction

Extension activities to use after the reproducible:

- Make the connection between equal parts of a whole and mathematical notations for fractions. Explain to students that the *denominator,* or bottom number, in fractions tells how many equal parts are in the whole object. The *numerator* tells how many of the parts are being used. To reinforce this concept, provide each student with five 9" x 2" construction paper strips. Then guide each student through the steps below. For more practice with fractions, have each student cut apart his strips along the folds and store the pieces in a large resealable plastic bag. Allow students to use these self-made manipulatives to explore other fraction-related concepts, such as naming fractions greater or less than $1/8$, finding fractions equivalent to $1/2$, or finding the difference between $6/8$ and $2/8$.

    Steps:
    1. Label the first strip "$1/1$" to show one whole.
    2. Fold the next strip in two equal parts. Label each part "$1/2$."
    3. Fold the third strip twice to make four equal parts. Label each part "$1/4$."
    4. Fold the fourth strip three times to make eight equal parts. Label each part "$1/8$."
    5. Fold the last strip four times to make 16 equal parts. Label each part "$1/16$."

- For a great literature connection, read aloud *Fraction Action* by Loreen Leedy. While reading, have student volunteers respond to Miss Prime's questions as they appear on pages 10, 15, 21, 27, and 31. Upon completion of the book, instruct each student to write his own fraction-related question on a 9" x 6" piece of construction paper and exchange papers with a classmate. Next, have each student read his classmate's question, turn the paper over, and then write the answer to the question and draw a matching illustration (if applicable). After all responses have been checked, assemble the pages into a class book titled "Room [room number]'s Fraction Action."

Equal parts of a whole

Name _____

Equal parts of a whole

Sailing Along With Fractions

Help keep this sailboat on the right course.
Study the figure in each box.
If the figure is divided into equal parts, color the box to show the course.

Word List
halves   thirds   fourths
fifths   eighths ninths

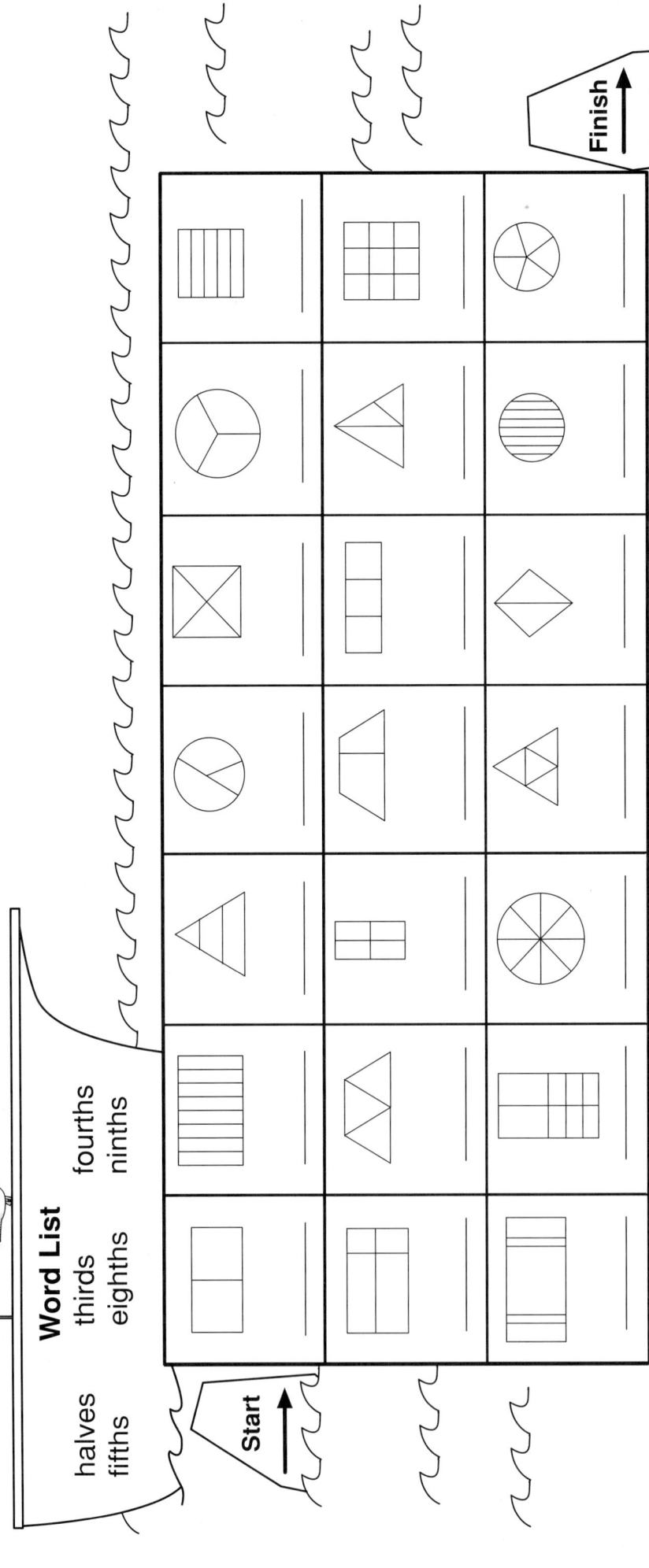

For each box on course, write the word that tells how the figure is divided.
Use the word list to help you with spelling.

Bonus Box: Draw four different figures. Divide one into halves, one into thirds, one into fourths, and one into sixths. Remember that the parts of each whole must be equal!

Getting a Jump on Fractions

Jump-start students' math skills with this fraction fun!

Purpose: To read and write fractions that describe parts of a group

Students will do the following:
- identify parts of a group
- write fractions that describe parts of a group

Materials for each student:
- copy of page 40
- pencil
- crayons

Vocabulary to review:
- fraction
- numerator
- denominator

Extension activities to use after the reproducible:
- This fraction activity is unbearably fun! Give each student a small paper plate with 12 edible teddy bear manipulatives. Under your direction, have students use the bears to solve challenges such as finding $\frac{1}{2}$ of 12, $\frac{1}{3}$ of 9, or $\frac{1}{5}$ of 10. Wrap up this sweet activity by having students eat their manipulatives.

- This fast-paced game gives students lots of practice identifying fractions of a group. Each small group of students needs a set of counters, a pair of dice, and access to a clock with a second hand. To play the game, each student rolls the dice in turn. The student has a predetermined number of seconds to decide to take $\frac{1}{2}$, $\frac{1}{3}$, or $\frac{1}{4}$ of the number shown on the dice in counters. (Have the player to the right of the student watch the clock.) If it is impossible for the student to take $\frac{1}{2}$, $\frac{1}{3}$, or $\frac{1}{4}$ of the number shown, his turn is over. Play continues until one player wins the game by collecting 20 counters.

Fractions of a group

Name _____ *Fractions of a group*

Getting a Jump on Fractions

Study each group of kangaroos.
Complete the fractions underneath to describe parts of each group.

A.

$\dfrac{\square}{3}$ = resting $\dfrac{\square}{3}$ = not resting

B.

$\dfrac{\square}{2}$ = digging $\dfrac{\square}{2}$ = not digging

C.

$\dfrac{\square}{4}$ = hopping $\dfrac{\square}{4}$ = not hopping

D.

$\dfrac{\square}{3}$ = eating $\dfrac{\square}{3}$ = not eating

E.

$\dfrac{\square}{5}$ = drinking $\dfrac{\square}{5}$ = not drinking

F.

$\dfrac{\square}{4}$ = babies $\dfrac{\square}{4}$ = adults

Bonus Box: Draw 12 kangaroos. Color $\frac{1}{2}$ of the kangaroos red, $\frac{1}{6}$ brown, and $\frac{1}{3}$ yellow.

Sweet Comparisons

Make comparing fractions a sweet treat for students!

Purpose: To compare fractions

Students will do the following:
- use pictures to solve problems
- color models to show fractions
- use mathematical symbols (>, <, =) to compare fractions

Materials for each student:
- copy of page 42
- pencil
- colored pencils

Vocabulary to review:
- compare
- greater than (>)
- less than (<)
- equal to (=)

Extension activities to use after the reproducible:

- Wrap up the activity on page 42 by having youngsters make fraction bars of their own. Provide each youngster with four 2" x 6" brown construction paper strips (chocolate bars), one 8" x 5" piece of foil, one 6" x 6" white construction paper square, and one jumbo paper clip. Each student folds one bar into halves, one into thirds, one into fourths, and one into eighths. She then labels the sections of each bar respectively. Next, she fastens the bars with the paper clip, wraps the foil around her fraction bars, and folds the foil ends toward the center. To make the outer wrapper, she centers her project atop the white paper, folds the top and bottom edges of the paper toward the center of the project, and then secures the overlapping edges with tape. Then she temporarily removes the fraction bars while she decorates and personalizes the resulting candy bar wrapper. Have each student use her sweet fraction bars to write and compare fractions of her own.

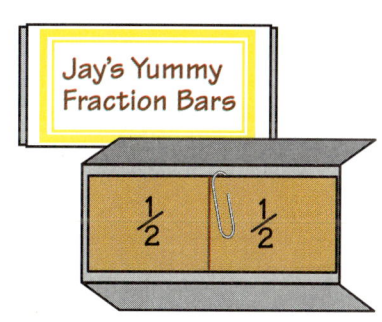

- Dominoes make perfect ready-made fraction manipulatives for this partner activity. Pair students and randomly give each pair five dominoes (see domino patterns on page 163). Point out to students how to read a domino as a fraction. Hold up a domino vertically with the face having fewer dots as the numerator and the face with more dots as the denominator. Then challenge each pair to order its dominoes from greatest to least according to the fractions they represent. Encourage youngsters to draw and shade pictures similar to those on page 42 to help them solve this challenge. Then instruct each pair to draw its final answer on a sheet of drawing paper.

Comparing fractions

Name _____ Comparing fractions

Sweet Comparisons

Color the candy bars in each pair to show each fraction.
Then compare the fractions.
Write >, <, or = in each circle.

THE GREAT "CANDO-OMATIC"

A.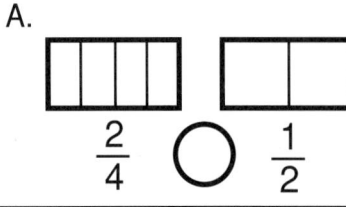
$\frac{2}{4}$ ◯ $\frac{1}{2}$

B.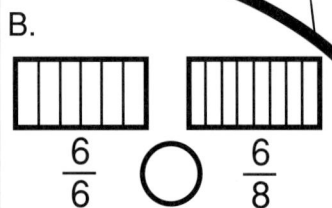
$\frac{6}{6}$ ◯ $\frac{6}{8}$

E.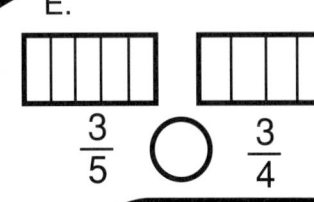
$\frac{3}{5}$ ◯ $\frac{3}{4}$

D.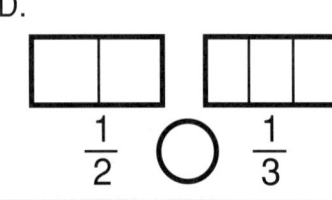
$\frac{1}{2}$ ◯ $\frac{1}{3}$

C.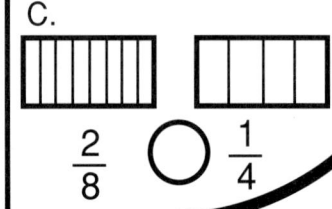
$\frac{2}{8}$ ◯ $\frac{1}{4}$

F.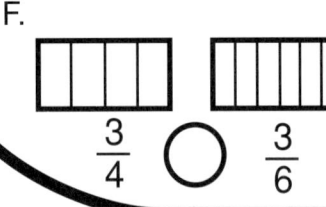
$\frac{3}{4}$ ◯ $\frac{3}{6}$

G.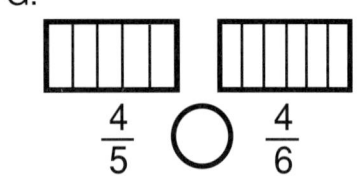
$\frac{4}{5}$ ◯ $\frac{4}{6}$

H.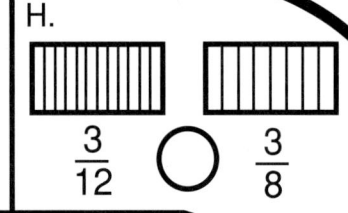
$\frac{3}{12}$ ◯ $\frac{3}{8}$

K.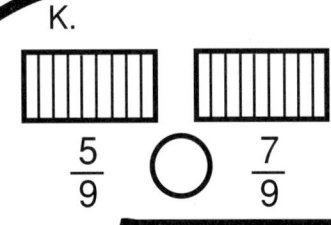
$\frac{5}{9}$ ◯ $\frac{7}{9}$

J.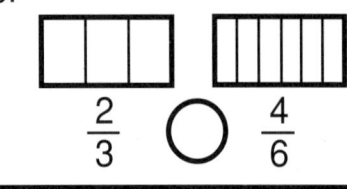
$\frac{2}{3}$ ◯ $\frac{4}{6}$

I.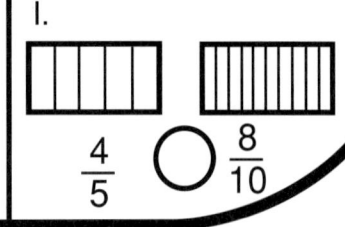
$\frac{4}{5}$ ◯ $\frac{8}{10}$

L.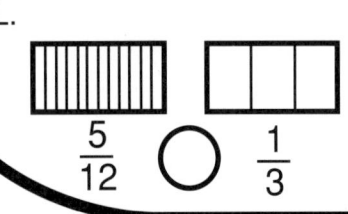
$\frac{5}{12}$ ◯ $\frac{1}{3}$

M.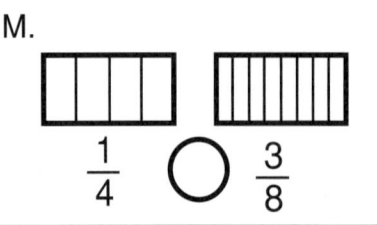
$\frac{1}{4}$ ◯ $\frac{3}{8}$

THE GREAT "WRAP-OMATIC"

Bonus Box: Draw a picture to solve the following problem: Fannie ran $\frac{1}{2}$ of a mile, Freddy ran $\frac{2}{3}$ of a mile, and Flossy ran $\frac{3}{4}$ of a mile. Who ran the farthest?

Putting a Spin on Fractions

Weave a web of understanding as students explore equivalent fractions!

Purpose: To investigate equivalent fractions

Students will do the following:
- identify fractional parts
- identify and create equivalent fractions

Materials for each student:
- copy of page 44
- pencil
- colored pencils

Vocabulary to review:
- numerator
- denominator
- equivalent

Extension activities to use after the reproducible:

- This activity for reinforcing equivalent fractions really measures up! Each small group of students needs a set of measuring cups. Copy the recipe shown on the board. Challenge each group to rewrite the recipe using equivalent fractions from the measuring cups. After each group finds the correct measures, allow it to mix a batch of the recipe and enjoy this tasty snack.

  $4/8$ c. Chex® cereal
  $2/8$ c. shelled peanuts
  $3/6$ c. Cheerios® cereal
  $2/6$ c. mini chocolate chips
  $3/12$ c. mini marshmallows
  $3/9$ c. raisins

- Students will take part in matching fractions with this concentration activity. Obtain 20 index cards. Program each pair of cards with pictures of equivalent fractions and write the corresponding fraction for each picture on the back of the card. Place the cards in a center. A student pair arranges the cards with the pictures facedown on a playing surface. Player 1 looks at each fraction, chooses an equivalent pair, and checks the picture on the back to see if she is correct. If she is correct, she gets to keep the cards. If she is incorrect, she returns the cards to their original positions. Player 2 takes a turn in the same manner. The game continues until all card pairs have been matched. The player with more pairs wins.

Equivalent fractions

Name _____ Equivalent fractions

Putting a Spin on Fractions

Color the circle on each spider to show the fraction.
Then find four pairs of spiders with equivalent fractions.
Use the same color to trace the draglines of the spiders in each pair.

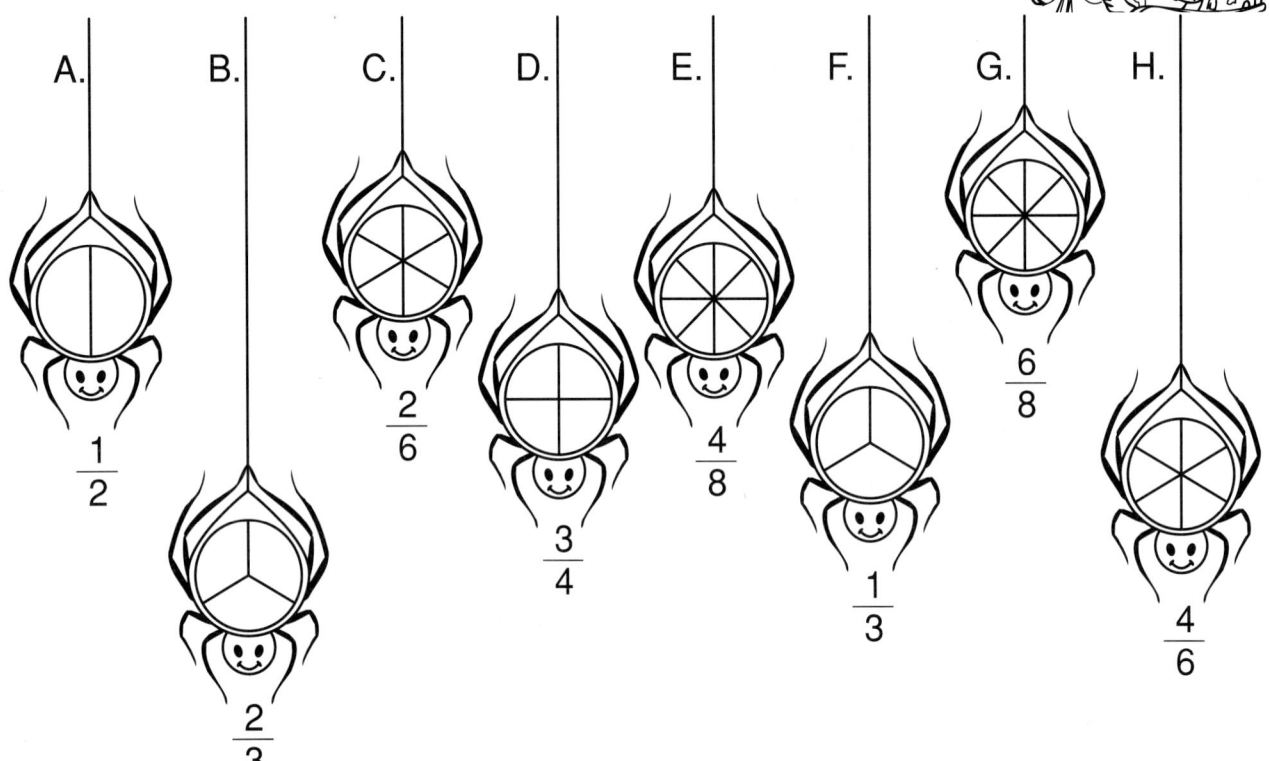

Color the spider to show an equivalent fraction for each.
Then write the numerator in the space provided.

I. $\frac{1}{4} = \frac{\square}{8}$

J. $\frac{1}{2} = \frac{\square}{4}$

K. $\frac{4}{4} = \frac{\square}{3}$

L. $\frac{1}{5} = \frac{\square}{10}$

M. $\frac{1}{6} = \frac{\square}{12}$

N. $\frac{3}{6} = \frac{\square}{2}$

Bonus Box: Write five equivalent fractions for $\frac{1}{2}$.

A Menu of Mixed Numbers

Increase students' appetites for fractions by giving them a taste of mixed numbers!

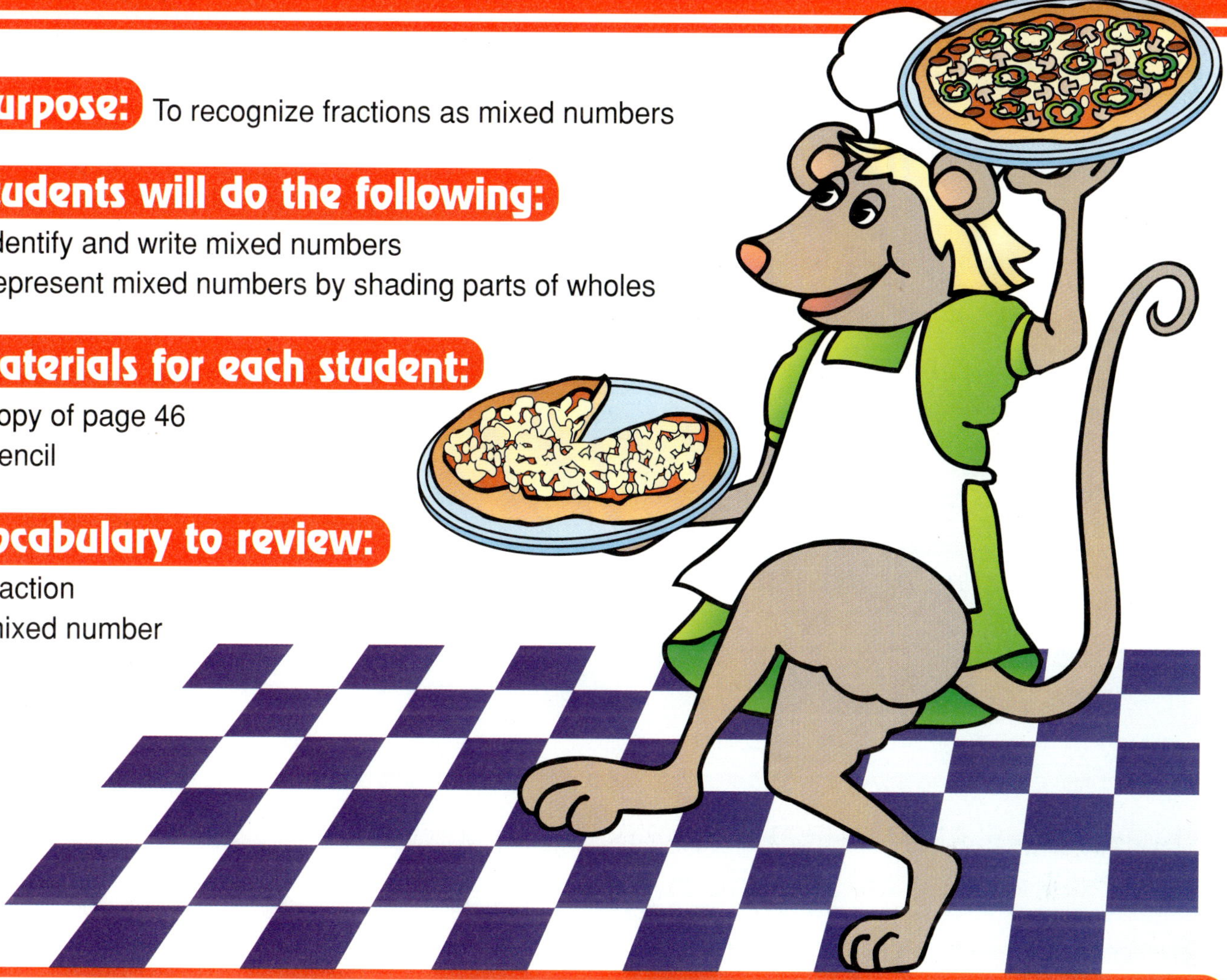

Purpose: To recognize fractions as mixed numbers

Students will do the following:
- identify and write mixed numbers
- represent mixed numbers by shading parts of wholes

Materials for each student:
- copy of page 46
- pencil

Vocabulary to review:
- fraction
- mixed number

Extension activities to use after the reproducible:

- Sweeten students' understanding of mixed numbers with graham crackers! Provide each pair of students with four graham crackers that can be divided into fourths and a paper towel. Instruct each pair to break each cracker's sections apart. Then have each twosome model mixed numbers such as $1\frac{3}{4}$, $2\frac{1}{2}$, $3\frac{1}{4}$, and $3\frac{2}{4}$ and check its model against a correct representation that you draw on the board. Then pose questions such as "If one-fourth of your crackers were eaten, how much would be left?" and "How many crackers is six-fourths?" For a tasty follow-up, reward each student with a fresh graham cracker to eat.

- This small-group activity is just perfect for reinforcing mixed numbers! Demonstrate the relationship between pattern blocks and fractions by presenting students with a key like the one shown. Provide each student with a copy of the pattern blocks on page 165. Have him color and cut out his pattern blocks and store them in a resealable plastic bag. Group students and give each group a mixed number cube programmed with the following mixed numbers: $1\frac{1}{3}$, $1\frac{1}{6}$, $1\frac{1}{2}$, $2\frac{1}{3}$, $2\frac{1}{6}$, and $2\frac{1}{2}$ (see cube pattern on page 162). To play, a student places his pattern blocks in sorted piles. In turn, he rolls the cube and uses his pattern blocks to model the number shown. Upon accumulation of blocks, he may exchange equivalent blocks for one whole. The first student to model five wholes or more wins the game.

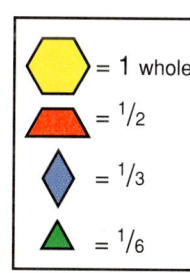

Mixed numbers

A Menu of Mixed Numbers

Name _____ Mixed numbers

Welcome to Myra's Pizza Buffet!
Write a mixed number to show how much pizza is on each tray.

1. [pizzas image]   2. 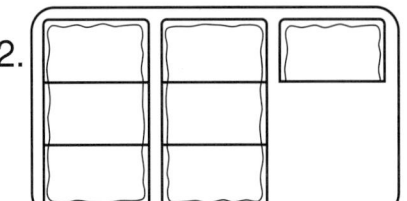   3.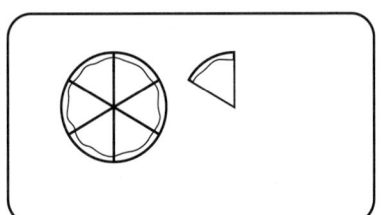

4. 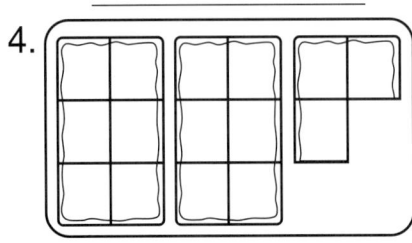   5. 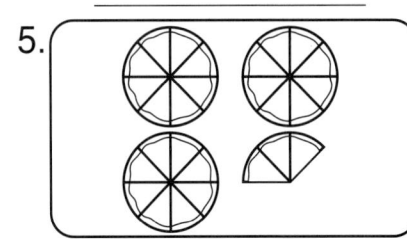   6.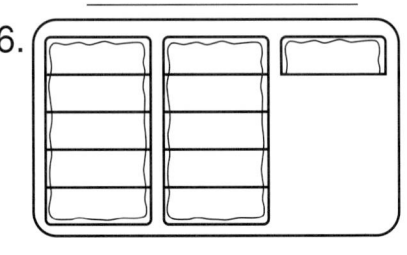

Read the amount of pizza each family ate.
Shade the pizzas to show the amount.

7. $2\frac{1}{2}$ pizzas
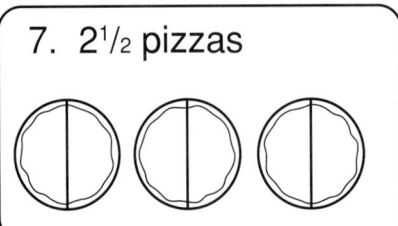

8. $1\frac{1}{4}$ pizzas
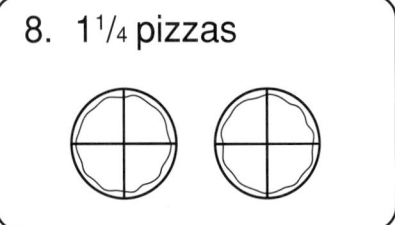

9. $2\frac{1}{3}$ pizzas
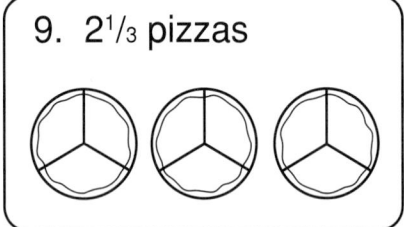

10. $3\frac{3}{8}$ pizzas
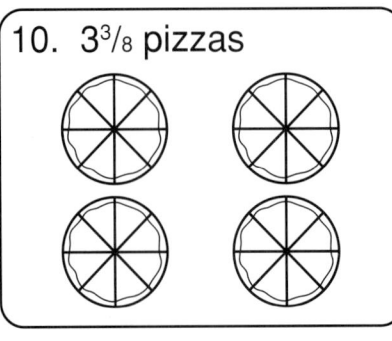

11. $4\frac{3}{4}$ pizzas
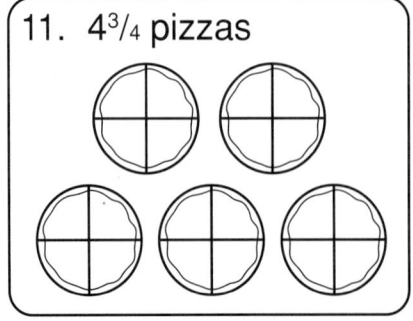

12. $3\frac{5}{6}$ pizzas
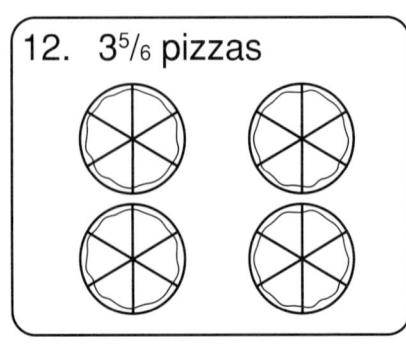

Bonus Box: Myra cuts pepperoni pizzas into eight slices. At the end of the day, she had $2\frac{1}{2}$ pepperoni pizzas. How many slices did she have left?

Playing a Round of Fractions and Decimals

Get your students into the swing of converting fractions and decimals!

Purpose: To learn how fractions and decimals are related

Students will do the following:
- relate fractions and decimals
- read and write decimal numbers to tenths

Materials for each student:
- copy of page 48
- pencil

Vocabulary to review:
- fraction
- decimal
- tenths

Extension activities to use after the reproducible:
- Have your youngsters get a taste of tenths with this sweet activity. Give each student ten M&M's® candies. Before students eat this tasty treat, have each student identify the number of tenths each color represents and write a corresponding decimal.

- Help students get a clear picture of hundredths decimals. Make and laminate a class supply of the hundreds chart on page 161. Make a transparency of the chart for yourself. Provide each student with a laminated chart, a wipe-off marker, and a paper towel to clean her chart. Write a hundredths decimal, such as .45, on the board. Have each student color the chart to represent the given amount and write the equivalent fraction. Then challenge each student to write the decimal and fraction for the remaining area. Shade the correct amount of squares on the transparency and allow students to check their work.

Fractions and decimals

Name_____   *Fractions and decimals*

Playing a Round of Fractions and Decimals

Write the equivalent fraction or decimal for each amount shown.

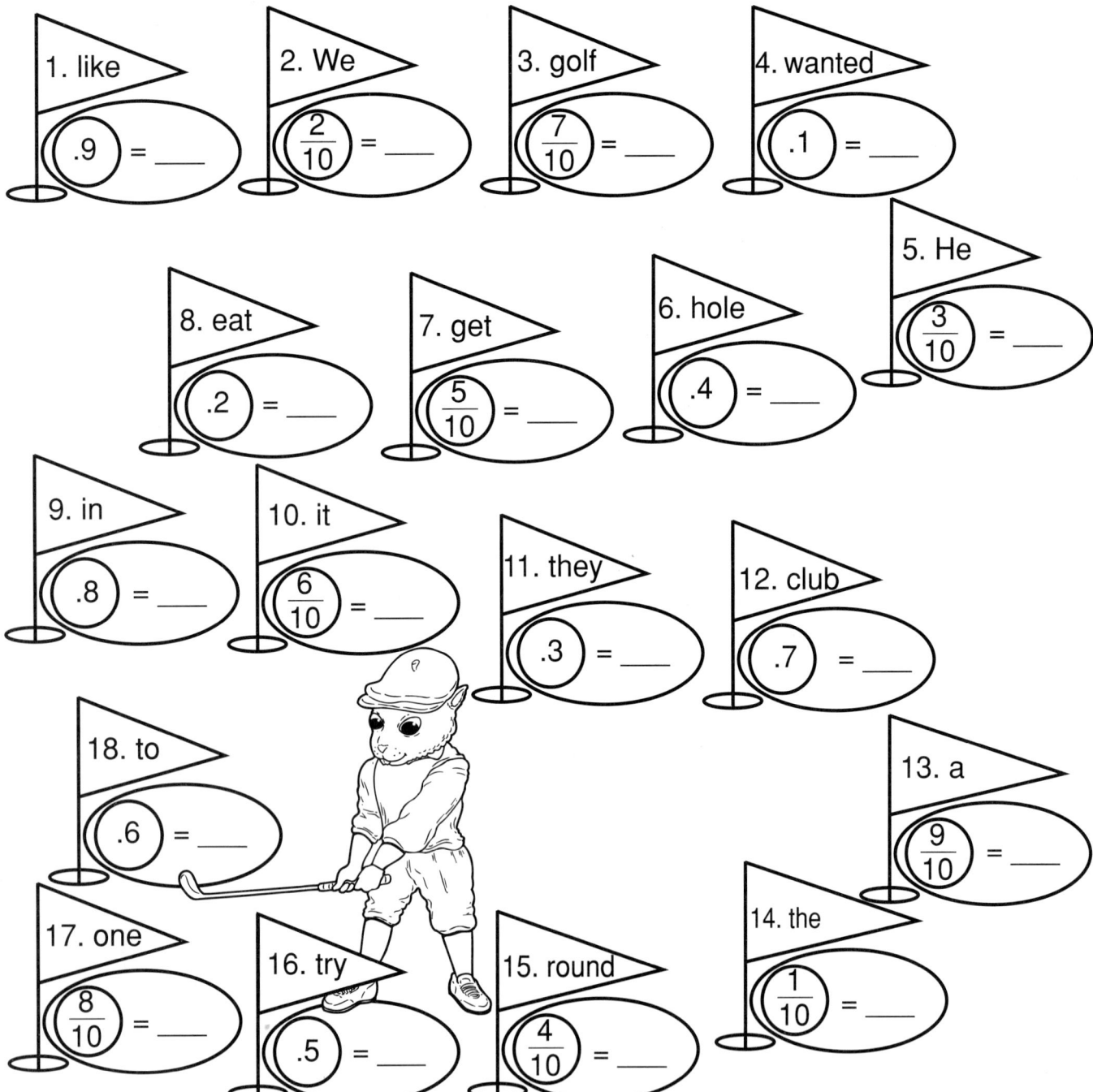

1. like .9 = ____
2. We 2/10 = ____
3. golf 7/10 = ____
4. wanted .1 = ____
5. He 3/10 = ____
6. hole .4 = ____
7. get 5/10 = ____
8. eat .2 = ____
9. in .8 = ____
10. it 6/10 = ____
11. they .3 = ____
12. club .7 = ____
13. a 9/10 = ____
14. the 1/10 = ____
15. round 4/10 = ____
16. try .5 = ____
17. one 8/10 = ____
18. to .6 = ____

Why did the golfer choose the doughnut for a snack?
To find out, write the matching word for each answer.

____ ____ ____ ____ ____ ____ ____ ____ ____ ____ !
 .3   1/10 6/10 .5   .9   4/10 8/10 .8

Bonus Box: Which of the following coins is one-tenth of a dollar: a penny, a nickel, a dime, or a quarter?

Snakes of All Sizes

Slither into nonstandard measurement with this "sss-uper" activity!

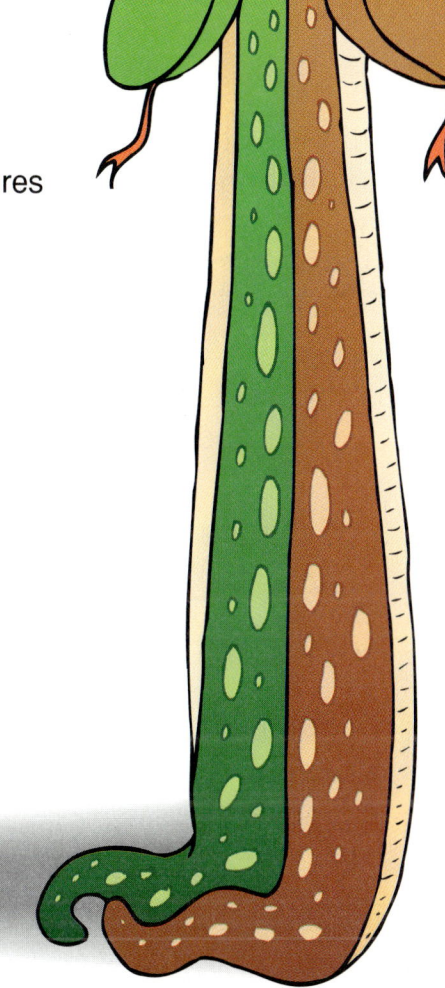

Purpose: To estimate and measure nonstandard units of length

Students will do the following:
- estimate length to the nearest nonstandard unit
- measure length using nonstandard units
- calculate the difference between estimates and actual measures

Materials for each student:
- copy of page 50
- pencil
- roll of Smarties® candies (if desired, provide an additional roll for a follow-up snack)

Vocabulary to review:
nonstandard
estimate
actual
length
difference

Extension activities to use after the reproducible:
- Link this hands-on activity to your study of nonstandard measurement. Distribute a random number of large paper clips to each student. Have each student link his paper clips to form a chain. Then challenge each student to quietly explore the classroom to find a variety of objects the same length as his chain. Have students record each finding. After several minutes, have students come together to share their discoveries. If desired, reward students who found ten or more objects.

- Read aloud *How Big Is a Foot?* by Rolf Myller to help youngsters understand that using nonstandard units of measurement can produce different results. Follow up the story by having each youngster trace his stockinged foot on a sheet of construction paper, cut out the tracing, and write his name on the resulting footprint. As a class, brainstorm a list of classroom items that could be measured using footprints. Then divide students into groups of three and assign each group an item on the list. Each group member uses his footprint to measure the item and records how his measurements compare to others in the group.

Nonstandard units of length

Name _____   Nonstandard units of length

Snakes of All Sizes

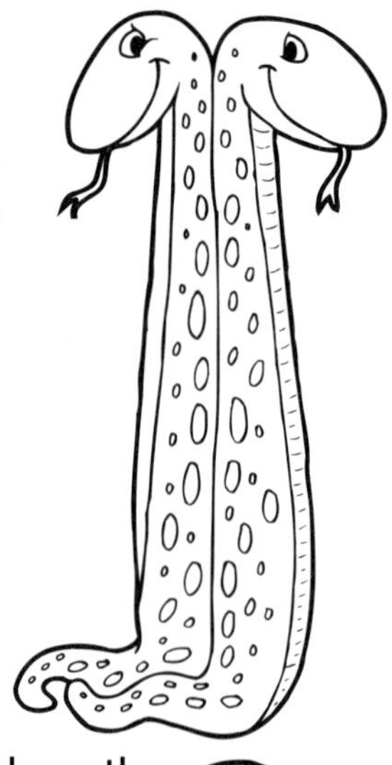

Measure these silly snakes before they slither away!
To practice measuring, place one of your candies over each circle.

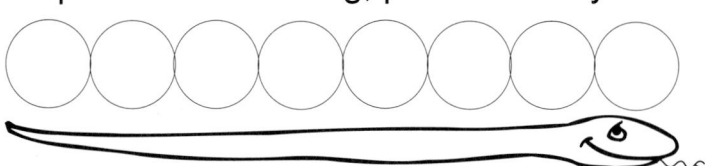

This snake is 8 candies long.

For each snake, estimate its candy length.
Then find the actual measure.
Write the difference between your estimated length and the actual length.

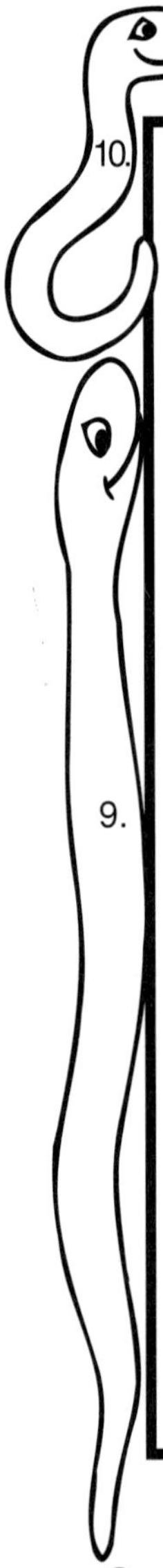

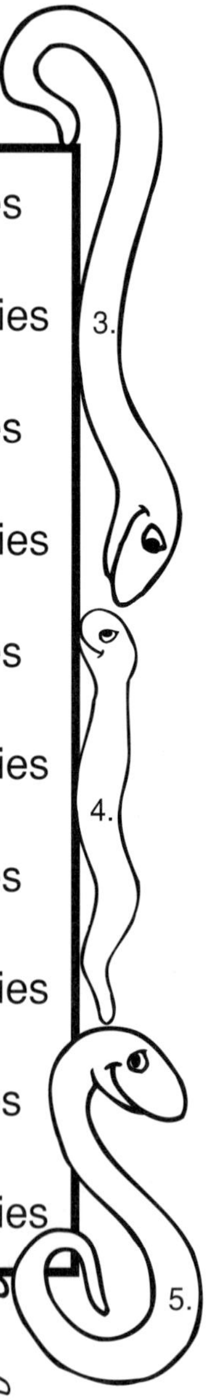

1. Estimate: _____ candies
   Actual: _____ candies
   Difference: _____ candies

2. Estimate: _____ candies
   Actual: _____ candies
   Difference: _____ candies

3. Estimate: _____ candies
   Actual: _____ candies
   Difference: _____ candies

4. Estimate: _____ candies
   Actual: _____ candies
   Difference: _____ candies

5. Estimate: _____ candies
   Actual: _____ candies
   Difference: _____ candies

6. Estimate: _____ candies
   Actual: _____ candies
   Difference: _____ candies

7. Estimate: _____ candies
   Actual: _____ candies
   Difference: _____ candies

8. Estimate: _____ candies
   Actual: _____ candies
   Difference: _____ candies

9. Estimate: _____ candies
   Actual: _____ candies
   Difference: _____ candies

10. Estimate: _____ candies
    Actual: _____ candies
    Difference: _____ candies

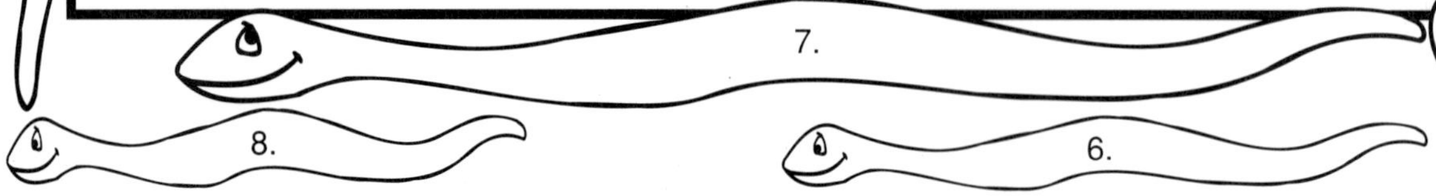

Bonus Box: Estimate how long all of your candies would be lying side by side. Draw a snake to match your estimate. Then measure with your candies to check your drawing.

Measuring Leaps and Bounds

Give students an inch...and they'll hop into measurement!

Purpose: To estimate and measure customary units of length

Students will do the following:
- estimate length to the nearest inch and half inch
- measure actual distance to the nearest inch and half inch

Materials for each student:
- copy of page 52
- pencil
- inch ruler (or a copy of the measuring strips on page 164)

Vocabulary to review:
- inch
- half inch

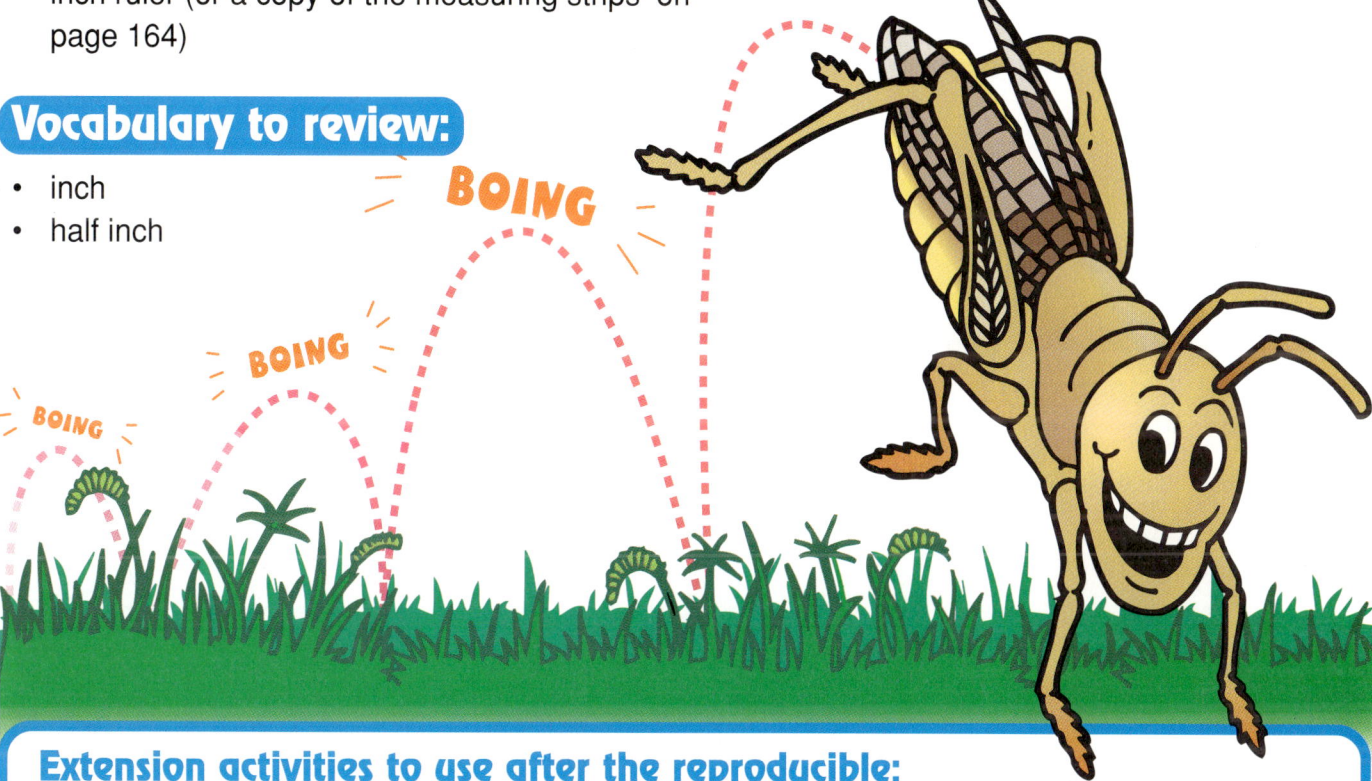

Extension activities to use after the reproducible:
- Hunting for a new measurement activity? Send students on a scavenger hunt! Pair students and provide each pair with a list of customary lengths and a ruler. Have each pair find a classroom object that matches each given length.

- Here's an easy-to-make learning center that offers students extra estimation and measurement practice. Gather a variety of items, such as a marker, a writing tablet, a crayon, and an eraser. For each item program an index card with its corresponding length. Place the items, programmed cards (facedown), and a ruler at a learning center. A student pulls a card, turns it faceup, and finds the corresponding object. After matching the object to the card, the student uses the ruler to check his guess before pulling the next card.

Customary units of length

Name _____ Customary units of length

Measuring Leaps and Bounds

This grasshopper can't seem to stay still!
First, estimate the distance of each of the grasshopper's leaps.
Then use a ruler to measure the actual distance between the ●s to the nearest half inch.
Record your answers on the lines.

The grasshopper leaps from the…	Estimate	Actual Measurement
1. leaf to the twig	_____	_____
2. twig to the dandelion	_____	_____
3. dandelion to the mushroom	_____	_____
4. mushroom to the pinecone	_____	_____
5. pinecone to the leaf	_____	_____
6. leaf to the rock	_____	_____
7. rock to the pinecone	_____	_____
8. pinecone to the twig	_____	_____
9. twig to the acorn	_____	_____
10. acorn to the mushroom	_____	_____

Bonus Box: If the grasshopper leaped from the rock to the pinecone, twig, leaf, acorn, dandelion, and mushroom, what would the total distance of his leaps be?

A Royal Ruler

Watch your youngsters measure with a ruler of a different kind!

Purpose: To estimate and measure metric units of length

Students will do the following:
- estimate length in centimeters
- measure length in centimeters
- identify reasonable units of measure

Materials for each student:
- copy of page 54
- pencil
- centimeter ruler (or a copy of the measuring strips on page 164)

Vocabulary to review:
- estimate
- centimeter
- meter

Extension activities to use after the reproducible:
- Build students' metric measurement skills by having them design a castle fit for a king. Provide each child with a piece of drawing paper and a metric ruler. Instruct each student to design and draw a castle. Set specifications for the width and height of any castle towers, windows, and doors. When the basic structure is completed, encourage each student to embellish his drawing with other details. If desired, display the completed projects on a bulletin board titled "Homes Fit for a King."

- Set students' metric measurement skills in motion! Have students assist you in determining a dance move to correspond with each of the following units of measurement: millimeter, centimeter, and meter. Then name various objects one at a time. Each student determines which unit of measurement is most reasonable for measuring the length of the object and demonstrates the corresponding dance move. Discuss the correct answer with students before naming the next object.

Metric units of length

Name _____   Metric units of length

A Royal Ruler

Estimate the length of each royal item in centimeters.
Then measure each item between the •s.
Write your answers on the lines.

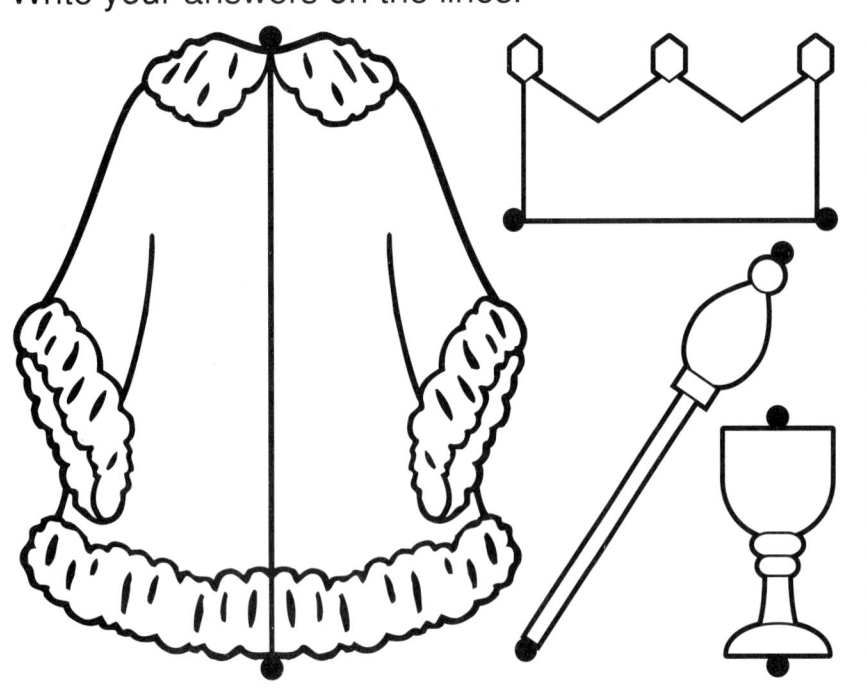

         Estimate Measurement   Estimate Measurement

1. robe _____ _____   4. cup _____ _____
2. crown _____ _____   5. flagpole _____ _____
3. staff _____ _____   6. chair _____ _____

How does my kingdom measure up?
Tell whether it is better to use centimeters
or meters to measure each item.

7. the drawbridge _____
8. the banquet table _____
9. the queen's necklace _____
10. the palace walls _____
11. the king's ruby ring _____
12. the king's smile _____

Bonus Box: List five items you could measure using centimeters and five items you could measure using meters.

A Weighty Decision

"Weight" and see how well your youngsters will learn to identify reasonable units of customary mass!

Purpose: To identify reasonable units of customary mass

Students will do the following:
- identify ounces or pounds as units of measurement
- identify reasonable amounts of ounces or pounds

Materials for each student:
- copy of page 56
- pencil

Vocabulary to review:
- ounces
- pounds

Extension activities to use after the reproducible:
- Combine estimation and measurement with this cooperative-group activity. Divide students into small groups. Provide each group with a scale that measures weight in ounces and a set of assorted items. Challenge each group to estimate the weight of each object and list the objects and estimates in order from lightest to heaviest. Then have each group weigh each item, record the actual measurement, and find the difference between the estimated and actual weight.

- A scale that measures weight in ounces is all you need for this classroom challenge. Divide students into small groups. Explain to students that objects such as a strawberry and a Ping-Pong® ball weigh about one ounce each, and objects such as a loaf of bread and a basketball weigh about one pound each. Lead students to understand that 16 ounces equals one pound. Then challenge each group to find and collect five or more classroom objects that total one pound. After a predetermined amount of time have each group weigh its collection of objects. The group closest to one pound wins the challenge.

Name _____   Customary units of mass

A Weighty Decision

Terry Tiger is gathering equipment for his sports camp.
Write ounces or pounds under each item.

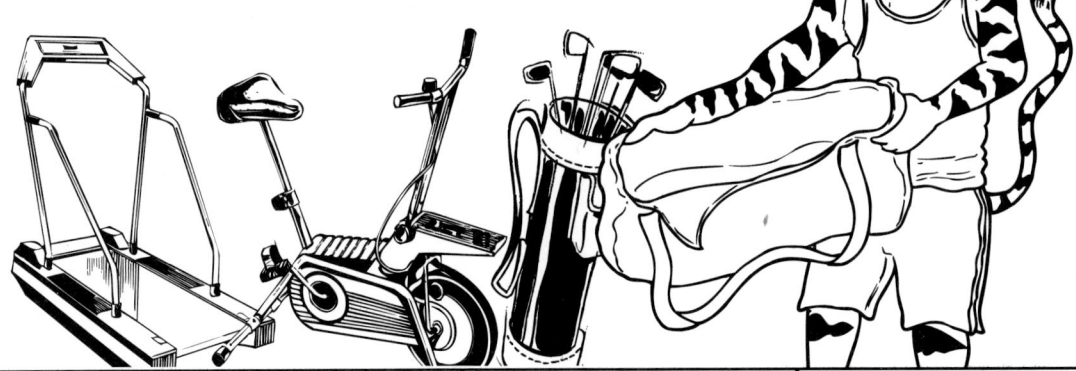

1. 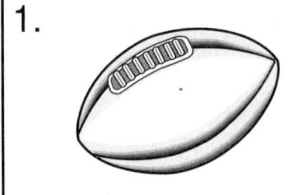 14 _____	2.  10 _____	3.  1 _____	4.  5 _____
5. 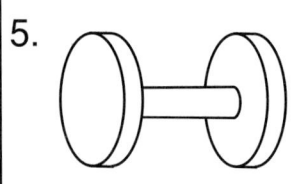 10 _____	6. 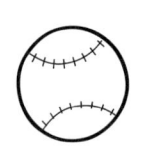 5 _____	7. 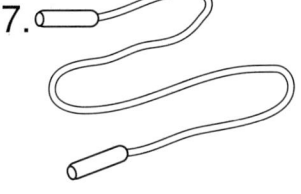 7 _____	8.  30 _____
9.  2 _____	10. 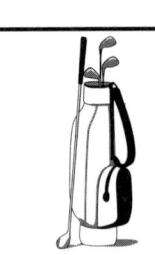 25 _____	11.  9 _____	12. 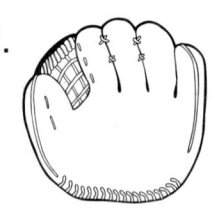 18 _____
13. 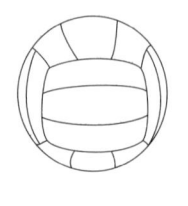 10 _____	14.  20 _____	15.  2 _____	**Bonus Box:** Terry's duffel bag will only hold 25 pounds. Color 10 things he can bring to the sports camp without going over 25 pounds. (Hint: 16 ounces = 1 pound)

56 ©2001 The Education Center, Inc. • Math Skills Workout • TEC3227 • Key p. 171

Jungle Journey

Take your youngsters on this journey to help them better understand grams and kilograms!

Purpose: To identify reasonable units of metric mass

Students will do the following:
- identify grams and kilograms as units of measurement
- identify reasonable amounts of grams or kilograms

Materials for each student:
- copy of page 58
- pencil

Vocabulary to review:
- mass
- gram (g)
- kilogram (kg)
- reasonable

Extension activities to use after the reproducible:

- This small-group competition will get your youngsters on a roll when it comes to distinguishing between grams and kilograms! Each small group of students will need a cube programmed as shown (pattern on page 162). Have each player list ten everyday items. Each player's list should include items that would be weighed better in grams, such as a pencil and a paper clip, and items that would be weighed better in kilograms, such as a person and a desk. To play, a student rolls the cube in turn and identifies one item on his list that would be better weighed using the unit shown. If all group members agree, the player crosses the item off his list. Play continues in this manner until one player wins the game by crossing off all of the items on his list.

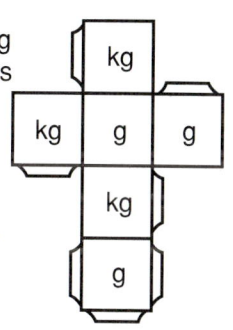

- Help students sort through metric units of mass in a flash. Give each student two index cards. Have each student write "grams" on the back of one card and "kilograms" on the back of the other. On the front of each card, each student illustrates and labels an object that corresponds to the metric mass shown. Check students' cards; then place corrected cards at a center illustration side up. A student looks at the picture on each card and places the card in one of two piles—items better weighed in grams or items better weighed in kilograms. Then he flips the cards in each pile to check his work.

Metric units of mass

Fishing for the Right Unit of Capacity

Hook students on customary capacity with this measurement activity!

Purpose: To understand customary units of capacity

Students will do the following:
- choose the most reasonable customary unit of capacity
- read a conversion table
- identify equivalent units of capacity

Materials for each student:
- copy of page 60
- pencil

Vocabulary to review:
- capacity
- cup
- pint
- quart
- gallon

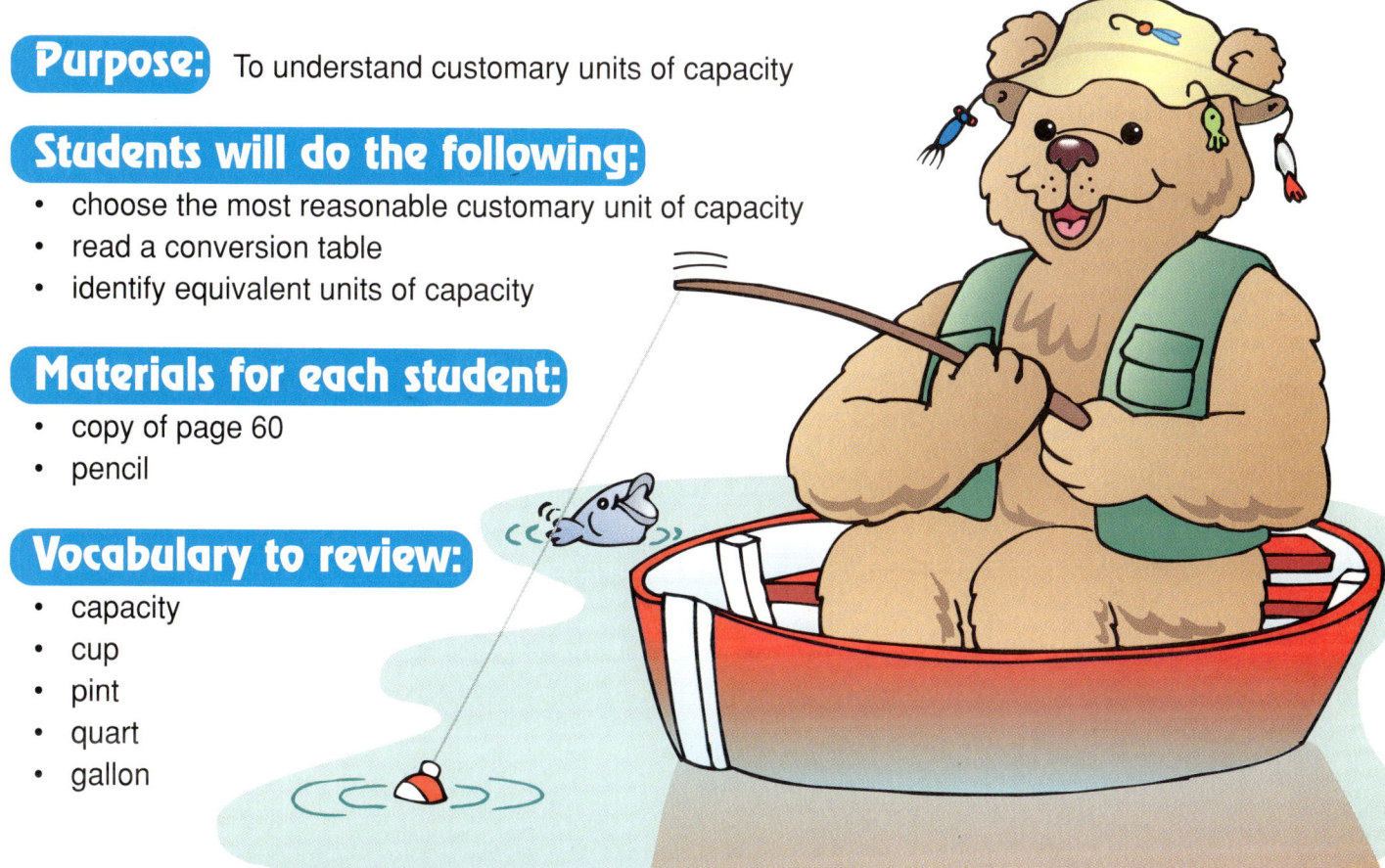

Extension activities to use after the reproducible:
- Get students into the swim of capacity by reading to them *Room for Ripley* by Stuart J. Murphy. After sharing the story, provide each student with a 9" x 12" sheet of light-colored construction paper. Instruct her to fold her paper in half twice and then unfold it to reveal four equal rectangles. Next, have her cut apart the four rectangles and divide three of them into labeled sections as shown. Provide each student with four sheets of duplicating paper. Have her glue a rectangle onto each sheet and then write an adventure similar to Carlos and Ana's in *Room for Ripley*, incorporating the four different units of capacity: gallon, quart, pint, and cup. Have her staple her four pages in order and then share her story with the rest of the class.

- Help students understand that containers with the same capacity can look very different. At a center, place a measuring cup, a container of dry beans, paper, and pencils. Also collect a variety of empty containers used to hold liquids, such as milk cartons, juice bottles, and soda bottles. Include two differently shaped containers for each unit of capacity—cup, pint, and quart—and label the containers 1–6. To use the center, partners select two containers that they estimate will hold equal amounts. One partner verifies each container's capacity by using the measuring cup and beans. If the containers have the same capacity, the other partner writes those matching numbers on a sheet of paper. If not, the steps are repeated with a different combination of containers. Students repeat this process with the remaining containers until all of the empty containers have been sorted into matching pairs.

Customary units of capacity

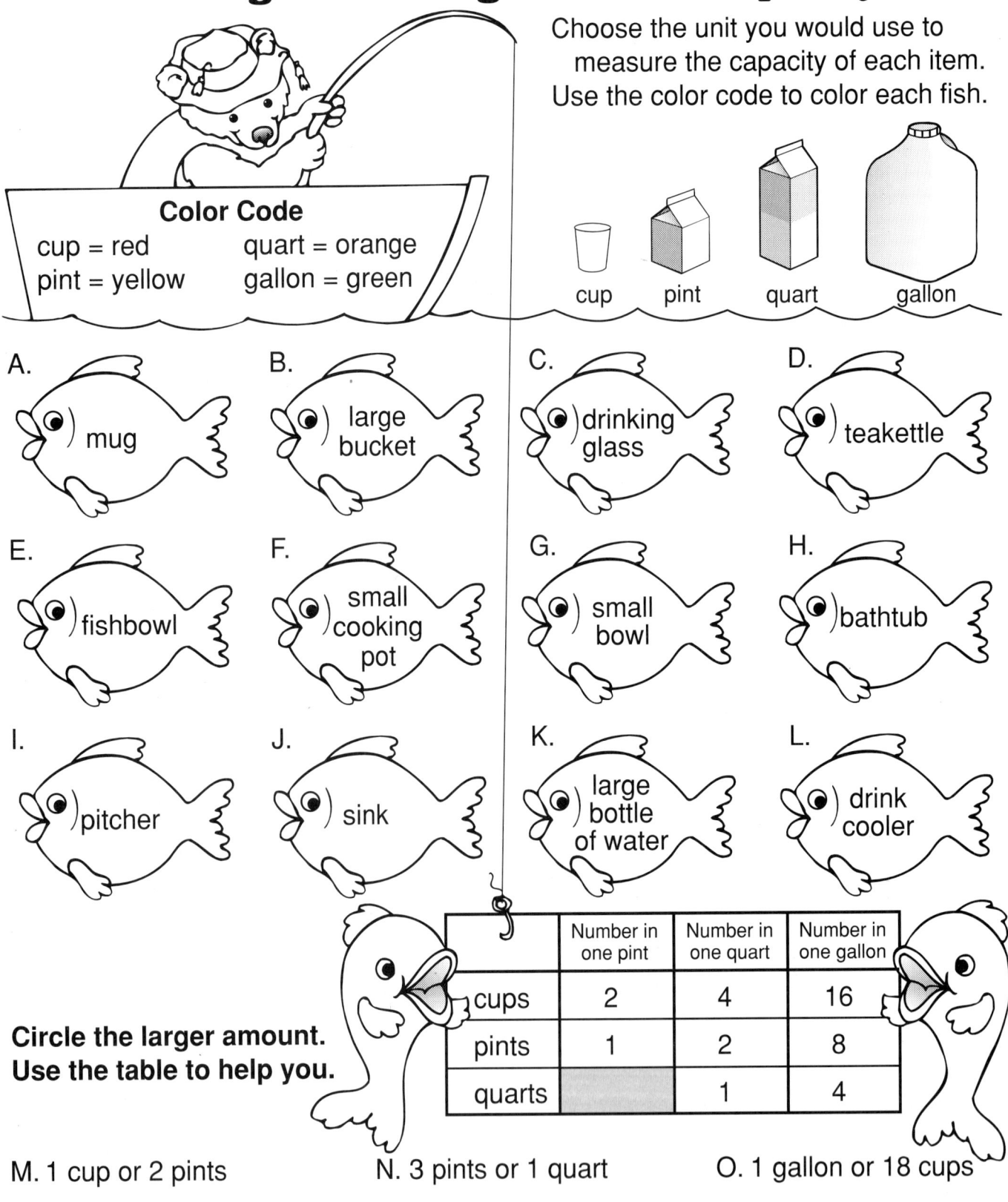

Ready to Travel?

Get students on the go with reading and understanding Fahrenheit temperature!

Purpose: To understand and measure Fahrenheit temperatures

Students will do the following:
- choose appropriate clothing for temperature ranges
- read thermometer models
- select appropriate activities for temperatures

Materials for each student:
- copy of page 62
- pencil

Vocabulary:
- Fahrenheit
- thermometer
- range

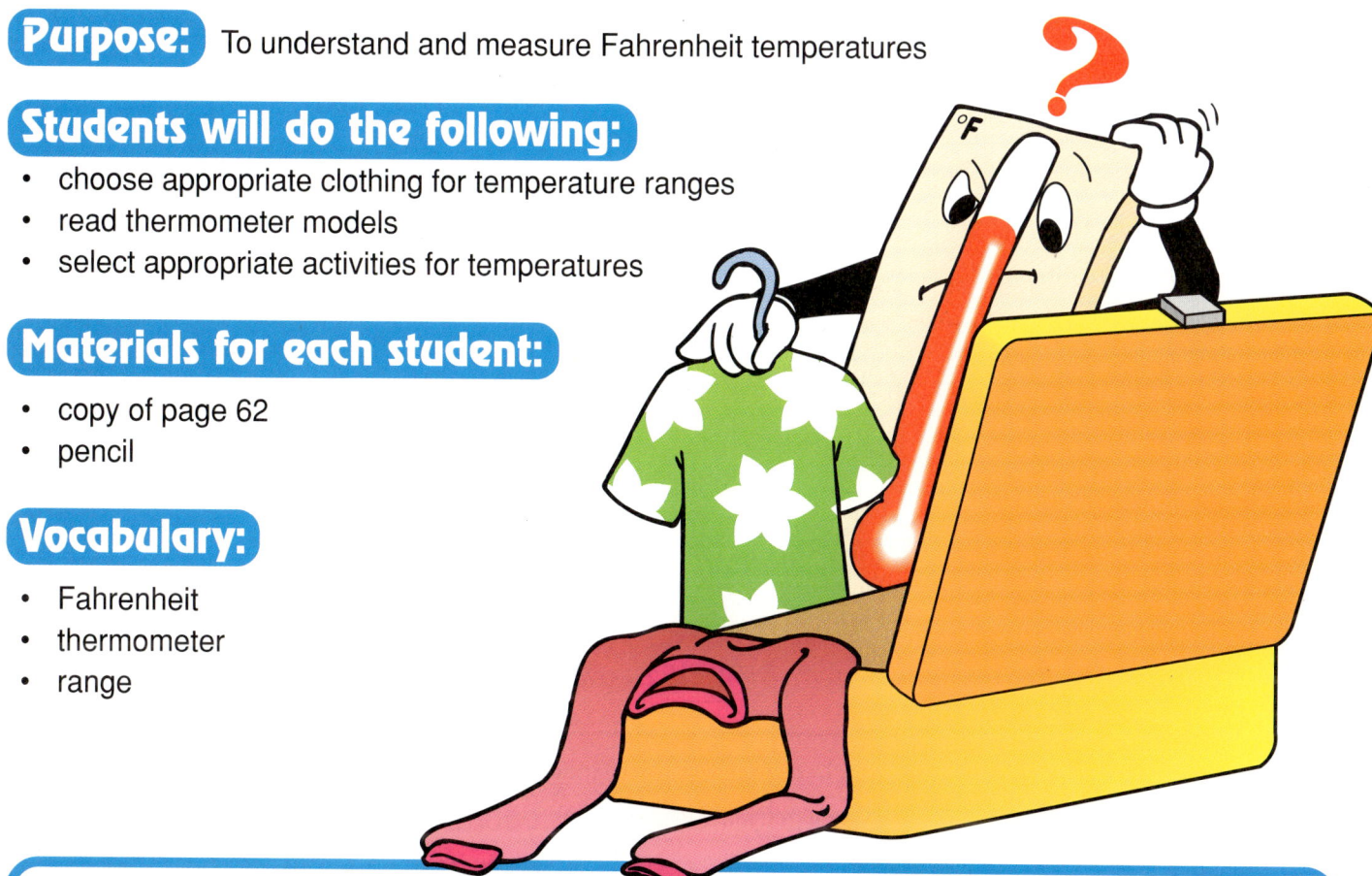

Extension activities to use after the reproducible:

- Assess students' understanding of Fahrenheit temperature and boost writing skills by having each student make a travel brochure. To make a brochure, each student folds a 9" x 12" sheet of construction paper into three equal sections and labels them "32°F," "70°F," and "90°F." Then he folds the sections closed to fashion a travel brochure. Next, each student thinks of a desired travel destination—such as the mountains, the beach, or a specific city—and illustrates and labels the location on the brochure's front flap. Then he opens the brochure. For each section, he writes about an activity visitors can do at the destination, based on the temperature shown, and draws a matching illustration.

- Use this problem-solving activity to give students practice reading thermometers. Give each student a construction paper copy of the thermometer pattern on page 164, a 1" x 14" strip of white construction paper, a red crayon, and scissors. Then have her make a thermometer according to the directions below. On the chalkboard, write a desired number of problems that give a series of directions, such as "Start with 27°F. Add 5°. Subtract 7°." Have students manipulate their thermometers to find each final temperature.

  Directions:
  1. Color half of the construction paper strip red.
  2. Use scissors to make a slit across each gray area on the thermometer.
  3. Slide the red end of the strip through the bottom slit. Then slide the white end through the top slit.
  4. Turn the thermometer over and tape the two ends together.

Fahrenheit temperature

Name _____   *Fahrenheit temperature*

Ready to Travel?

Read the temperature range on each suitcase.
Circle two items you probably would *not* need to pack.

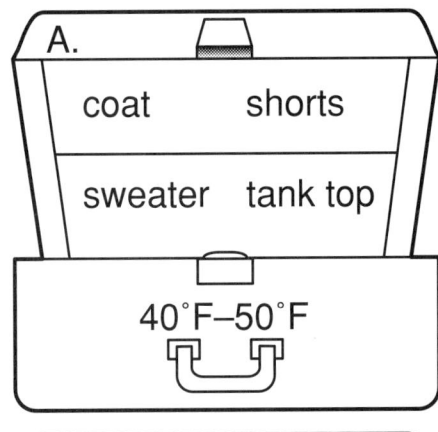
A. coat   shorts   sweater   tank top   40°F–50°F

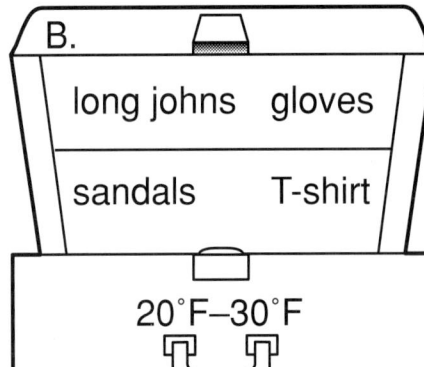
B. long johns   gloves   sandals   T-shirt   20°F–30°F

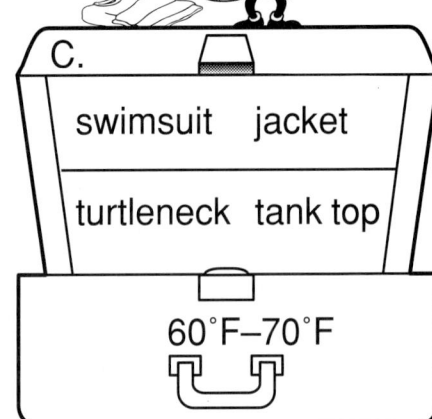
C. swimsuit   jacket   turtleneck   tank top   60°F–70°F

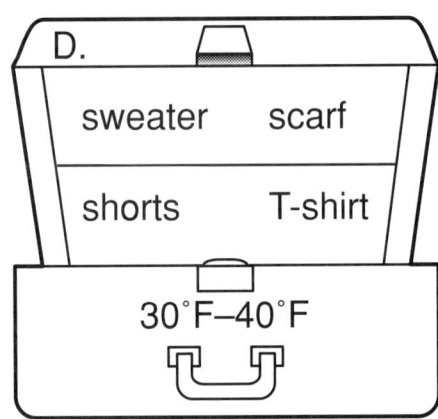
D. sweater   scarf   shorts   T-shirt   30°F–40°F

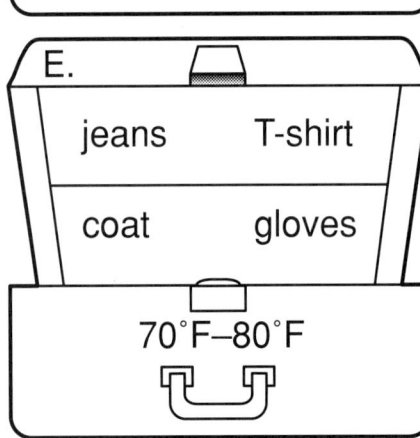
E. jeans   T-shirt   coat   gloves   70°F–80°F

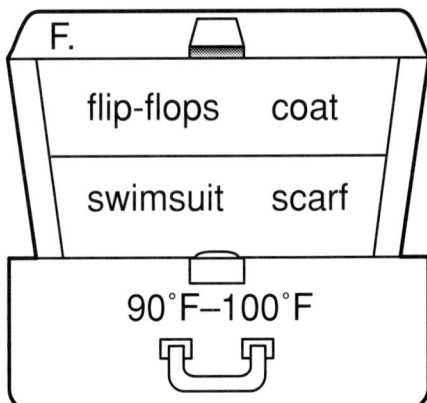
F. flip-flops   coat   swimsuit   scarf   90°F–100°F

Write the temperature shown under each thermometer.
Then circle the most reasonable outdoor activity for each temperature.

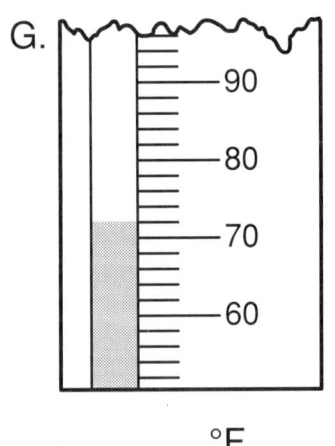

G. ____°F
fishing
skiing
swimming

H. ____°F
ice-skating
roller skating
horseback riding

I. ____°F
hiking
sledding
waterskiing

J. ____°F
snowboarding
jogging
swimming

Bonus Box: Choose a temperature shown on one of the thermometers. Write a story that takes place in a setting with that temperature.

Celsius Sense

Help students make sense of Celsius temperature while reinforcing thermometer reading skills!

Purpose: To understand and measure Celsius temperatures

Students will do the following:
- read Celsius thermometer models
- write temperatures in order from coldest to warmest
- create thermometer models to show given temperatures
- select reasonable activities for temperatures

Materials for each student:
- copy of page 64
- pencil

Vocabulary:
- Celsius
- thermometer
- reasonable

Extension activities to use after the reproducible:

- Try this instrument-making activity to strengthen students' understanding of Celsius temperature! Give each student a construction paper copy of the thermometer pattern on page 164 and a 1" x 14" strip of white construction paper. To make a thermometer, each student cuts a slit across each gray area on the thermometer pattern. Then she colors half of the white construction paper strip red, slides the red end of the strip through the bottom slit, and slides the white end through the top slit. Finally, she turns the thermometer over and tapes the two ends together. Pose questions to students such as "Which temperature is most appropriate for sledding: 0°C, 20°C, or 30°C?" (*0°C*). Have students manipulate their thermometers to display each answer.

- Track down a variety of Celsius readings in and around your school. In advance, invite several parent volunteers to assist with this activity. Assign each parent volunteer a small group of students. Have the parent volunteer lead the group in brainstorming a list of five sites that may have differing temperatures, such as the cafeteria, the principal's office, the gym, the hallway, or the playground. Give each group a Celsius thermometer with which to record the temperature of each location. Each parent volunteer guides his group to each site, has group members read and agree on the temperature, and has a group member record the findings. Upon returning to the classroom, have parent volunteers guide each group in making a bar graph to show its findings.

Celsius temperature

Name _____   Celsius temperature

Celsius Sense

Write each temperature shown.
Cross out each temperature on the list as you use it.

Temperature List
28°C
25°C
40°C
15°C
−15°C
35°C
−23°C
7°C
37°C
32°C
−30°C

A. 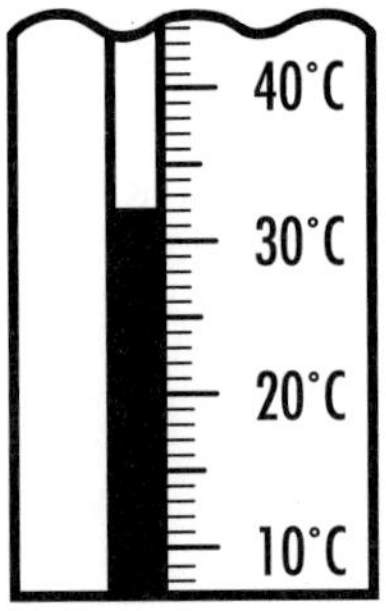 ___°C B. 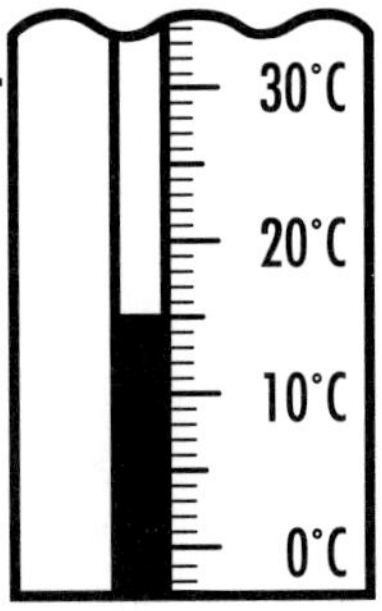 ___°C C. 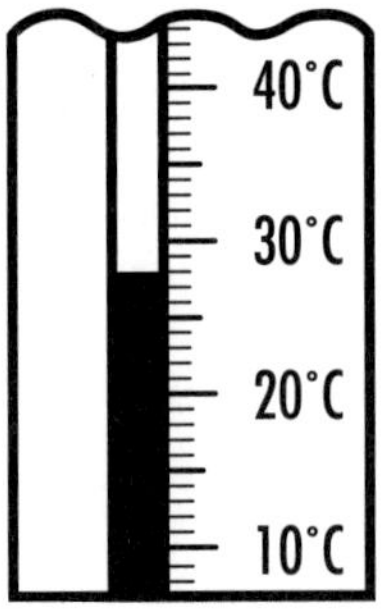 ___°C

D. 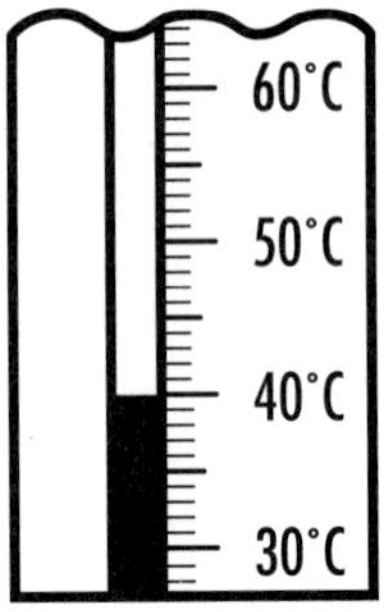 ___°C E. 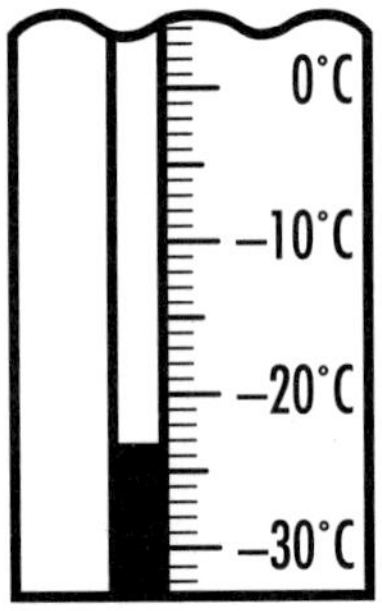 ___°C F. 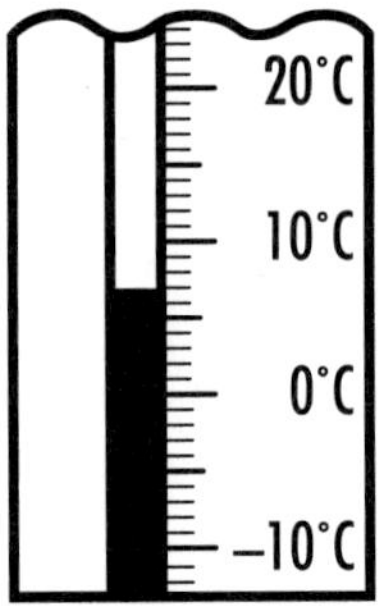 ___°C G. 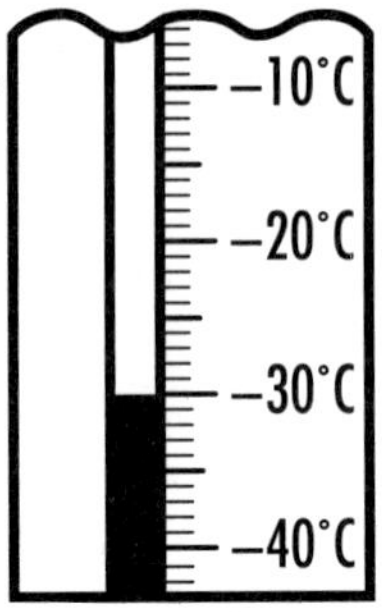 ___°C H. 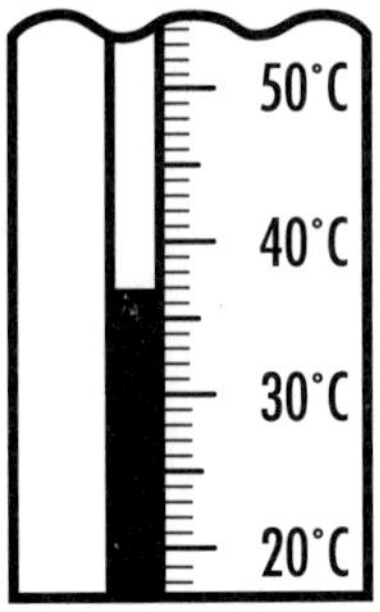 ___°C

Under the thermometers, write the temperatures you did not cross out in order from coldest to warmest.
Shade each thermometer to show each temperature.
Then choose a reasonable activity on the list for each temperature.
List each activity under the correct temperature.

Activity List
swimming
sledding
gardening

I. 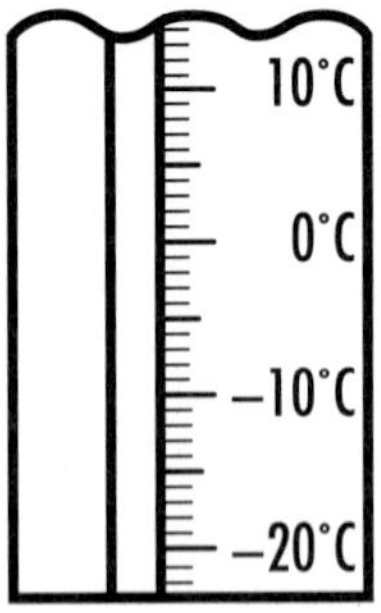 ___°C J. 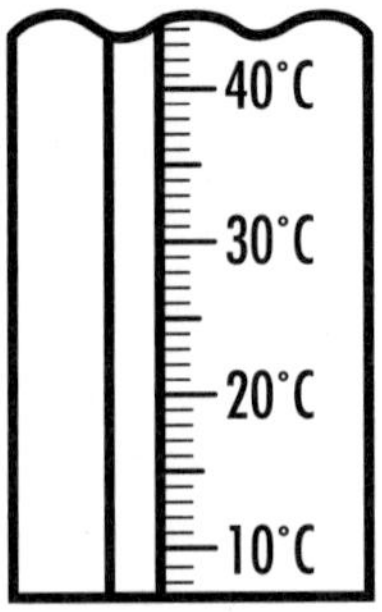 ___°C K. 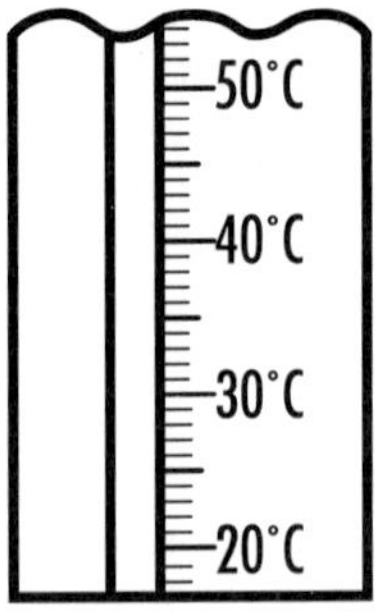 ___°C

Bonus Box: Which is your favorite temperature: 35°C or 10°C? Give three reasons why the temperature is your favorite.

A Perimeter Puzzler

Give students plenty of perimeter practice with these playful puzzle pieces!

Purpose: To calculate perimeter

Students will do the following:
- measure the sides of a figure to the nearest centimeter
- calculate the perimeter of a figure
- assemble a puzzle to form a rectangle

Materials for each student:
- copy of page 66
- pencil
- metric ruler (or a copy of the measuring strips on page 164)
- crayons
- scissors
- glue

Vocabulary to review:
- perimeter
- centimeter

Extension activities to use after the reproducible:

- Combine the concepts of perimeter and nonstandard measurement. In advance, cut a supply of construction paper rectangles of various sizes for each small group of students. Provide each group with the rectangles, ten large paper clips, and 15 small paper clips. Instruct each group to use the large paper clips to measure each side of a chosen rectangle; then have the groups add each length to find the perimeter. Next, encourage each group to estimate the perimeter of the same rectangle in small paper clips. Guide students to discover that the perimeter will be a greater number because it takes more small paper clips to cover the perimeter than large paper clips. Have each group measure the remaining rectangles in this manner.

- Extend students' understanding of perimeter with this hands-on algebraic task. Provide each student with a 30-centimeter pipe cleaner and a ruler. Instruct each student to bend her pipe cleaner into a triangle, place it on a sheet of paper, and then measure and label two sides. Explain to students that each pipe cleaner is 30 centimeters. Then have each student use a desired problem-solving strategy to find the missing length of her triangle without measuring. Allow each student to measure the length to verify her answer. If desired, encourage youngsters to illustrate their shapes and write equations as shown. Challenge students to repeat this process to find missing lengths of different shapes.

Perimeter 65

Name _____ Perimeter

A Perimeter Puzzler

Find the perimeter for each puzzle piece.
Color and cut out each piece.
Arrange the pieces to fit in the rectangle below.
Glue.

Bonus Box: Become a perimeter puzzler superstar! Write the perimeter of the complete puzzle in the ☆.

©2001 The Education Center, Inc. • *Math Skills Workout* • TEC3227 • Key p. 172

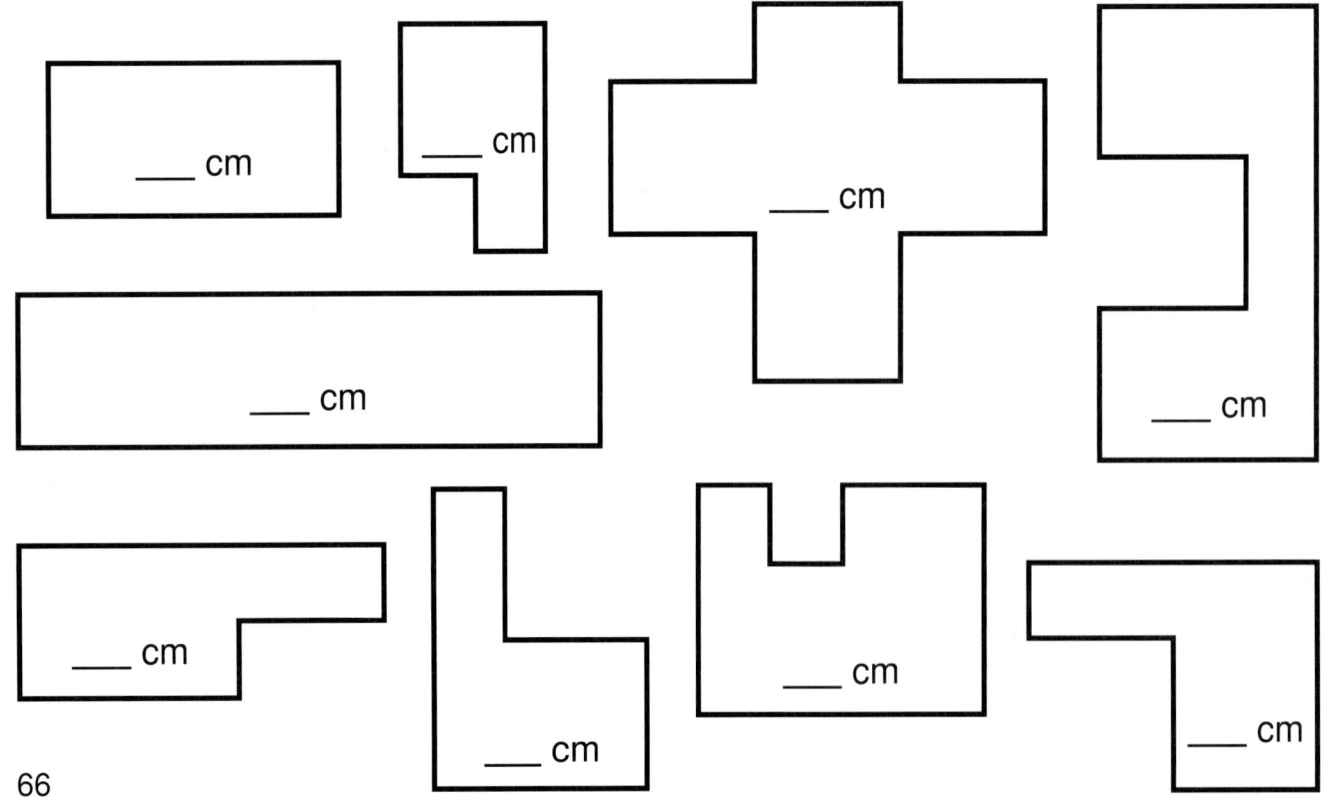

Sammy's Sports Complex

Shape up students' skills with finding area!

Purpose: To find the area of rectangular figures

Students will do the following:
- calculate the area of rectangular figures in centimeters
- determine the formula for area
- apply the formula for area to solve problems

Materials for each student:
- copy of page 68
- pencil

Vocabulary to review:
- area
- length
- width
- square centimeter

Extension activities to use after the reproducible:

- Reinforce area and logical reasoning with pentominoes! Show a square tile to the students. Point out that the tile represents one square unit. Ask students to predict the area of a figure with five tiles *(five square units)*. Then provide each student with five tiles and a copy of the centimeter graph paper on page 167. Instruct each student to create 12 different figures with an area of five square units by arranging the tiles so that at least one edge of each tile is touching the edge of another tile. (See possible answers below.) Have him shade squares on the graph paper to document each arrangement. Remind students that a flip or reversal of an arrangement does not count as a new arrangement. Guide students to understand that figures with different shapes can have the same area.

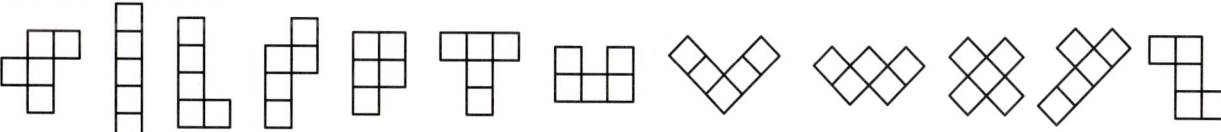

- Here's the perfect formula for connecting area and multiplication! Give each student pair 24 square tiles. Challenge the pair to find all the possible ways to make a rectangle of a given area. For example, if the area is 12, the twosome could assemble three rectangles with dimensions 1 x 12, 2 x 6, and 3 x 4. After assembling each rectangle, have the pair sketch each one on drawing paper, label the length of the sides, and write a corresponding multiplication sentence.

Area   67

Sammy's Sports Complex

Look at the floor plan for Sammy's Sports Complex. Write the length and width of each activity zone. Then count the squares to find the area. Write your answers on the chart. The first one has been done for you.

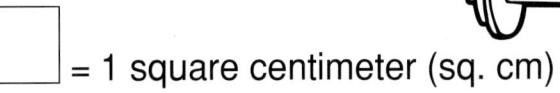 = 1 square centimeter (sq. cm)

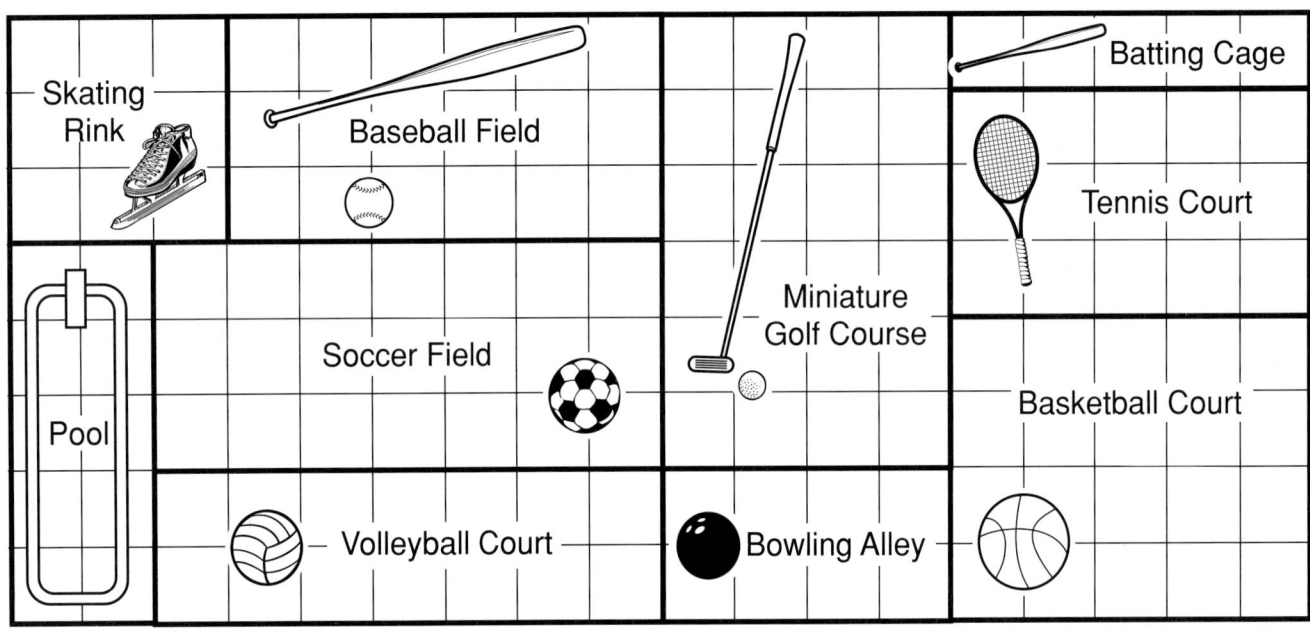

Activity Zone	Length	Width	Area
Skating Rink	3 cm	3 cm	9 sq. cm
Baseball Field			
Miniature Golf Course			
Batting Cage			
Tennis Court			
Pool			
Soccer Field			
Volleyball Court			
Bowling Alley			
Basketball Court			

Now answer each question.

1. Sammy wants to add a gymnastics area to his complex with a length of 6 cm and a width of 5 cm. What is the area? _____

2. What is the total area covered by all ten activity zones? _____

3. Study the chart. Look at the length, width, and area for each zone. What pattern do you see? _____

Bonus Box: The next set of floor plans shows a square-shaped locker room with an area of 16 sq. cm. Draw the plans for the locker room.

Under Construction

Build students' understanding of volume!

Purpose: To identify the volume of solid figures

Students will do the following:
- apply spatial sense to construct models of solid figures
- identify volume by counting cubic units

Materials for each student:
- copy of page 70
- pencil
- 30 interlocking cubes

Vocabulary to review:
- volume
- cubic units
- length
- width
- height

Extension activities to use after the reproducible:

- Lead students to discover the trick for finding volume of rectangular prisms! Pair students and give each pair a supply of interlocking cubes. Have each twosome build a rectangular prism and hide it from your view. Tell students that you will magically discover the volume of their prisms by asking three questions. Draw a chart on the chalkboard like the one shown. Ask a pair the length, width, and height of its prism and record it on the chart. Then, with much fanfare, write the volume of the figure on the chart. Have pairs count their cubes to verify your answer. Repeat this process with each pair of students. Next, ask students to look for a pattern to discover your magic trick before revealing that the product of the length, width, and height equals the volume of a rectangular prism.

Length	Width	Height	Volume

- After youngsters understand how to find the volume of a rectangular prism by multiplying the length, width, and height of the figure, use this learning center idea to put their problem-solving skills to the test. Program ten index cards with challenges similar to the ones shown. Then program the back of each card with the correct answer. Store the cards and a set of interlocking cubes at a center. A student reads each challenge, builds the figure to answer each question, and then flips the card to check her work.

Build a figure with a volume of 12 cubic units, a length of 3 units, and a height of 2 units. What is the width of the figure?

Build a figure with a volume of 27 units and the same length, width, and height. How many units are the length, width, and height?

Finding volume

Name _____   *Finding volume*

Under Construction

Build each figure with your cubes.
Write the volume in cubic units in the space provided.

Each ▢ = 1 cubic unit

1.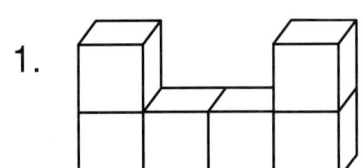

    _____ cubic units

2.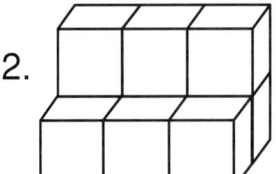

    _____ cubic units

3.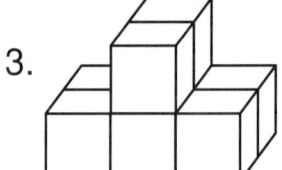

    _____ cubic units

4.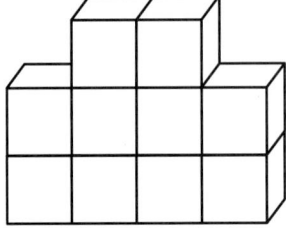

    _____ cubic units

5.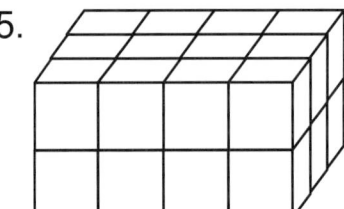

    _____ cubic units

6.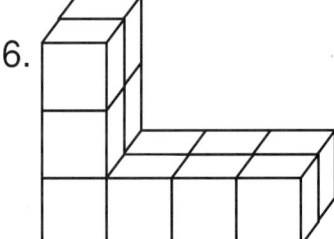

    _____ cubic units

7.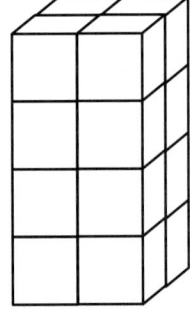

    _____ cubic units

8.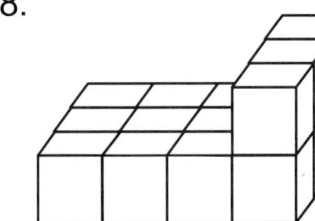

    _____ cubic units

9.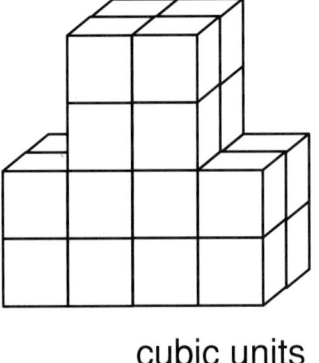

    _____ cubic units

10.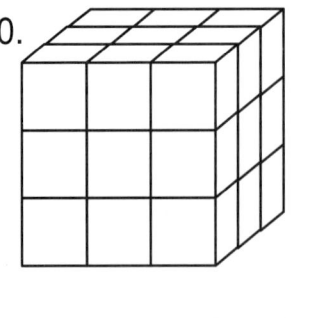

    _____ cubic units

Bonus Box: Look at the figures above. Circle each figure that is a rectangular prism or cube.

70   ©2001 The Education Center, Inc. • *Math Skills Workout* • TEC3227 • Key p. 172

Cuckoo for Clocks!

Watch your youngsters tell time like clockwork!

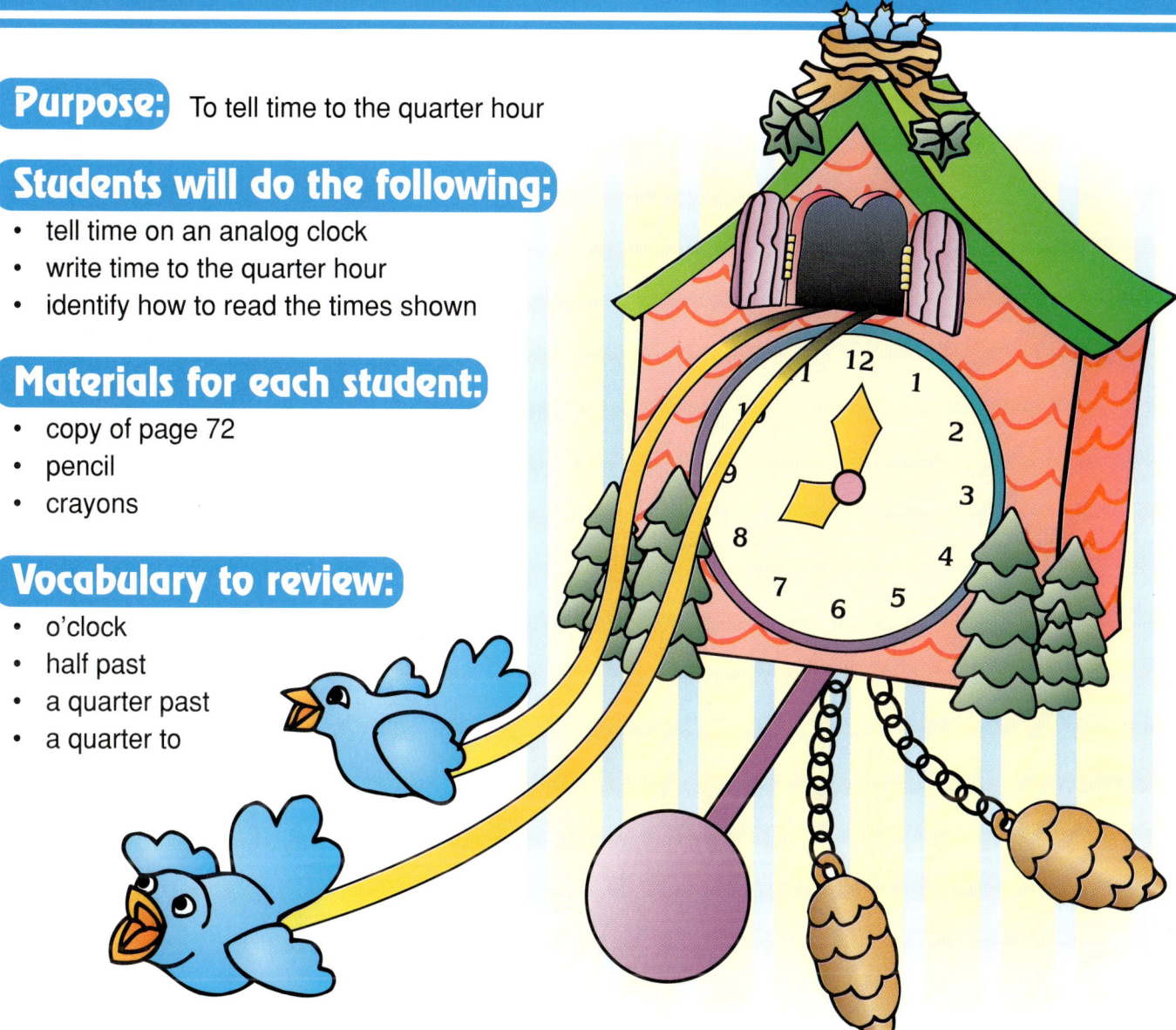

Purpose: To tell time to the quarter hour

Students will do the following:
- tell time on an analog clock
- write time to the quarter hour
- identify how to read the times shown

Materials for each student:
- copy of page 72
- pencil
- crayons

Vocabulary to review:
- o'clock
- half past
- a quarter past
- a quarter to

Extension activities to use after the reproducible:
- Have students tell time throughout the day with this activity. Provide each student with a copy of a table similar to the one shown. During the day, have each student record the start and end time for each activity. Then, when students are ready to practice elapsed time, they will already have personalized problems to solve!

- Take time out for this literature-based activity! Provide each student with a manipulative clock (or use copies of the clock pattern on page 164). Read the story *Clocks and More Clocks* by Pat Hutchins. As you read, invite students to manipulate their clocks to show each time mentioned in the story. Then have each student retell a similar story to a partner using his clock to help him.

Time to the quarter hour 71

Name _____ Time to the quarter hour

Cuckoo for Clocks!

For each clock, write the time shown.
Think about words you would use to read the times.
Then use the code to color each clock.

A.

B.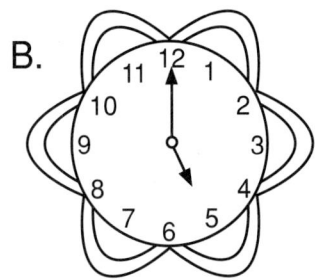

C.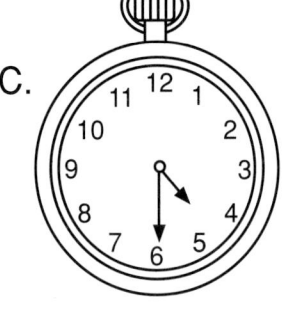

D.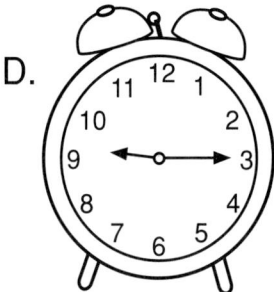

E.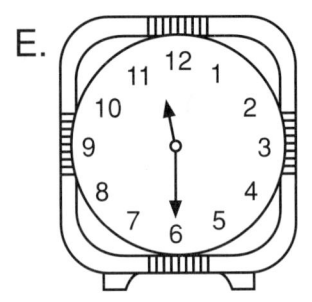

F.

G.

H.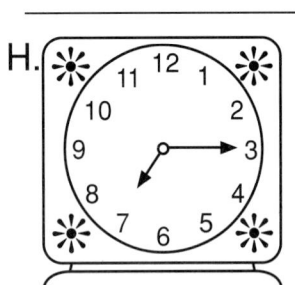

I.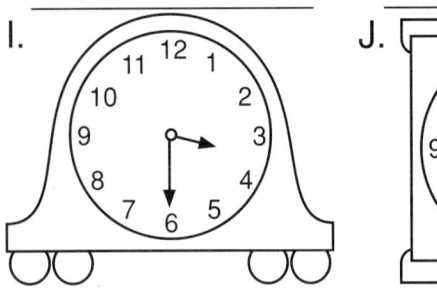

J.

K.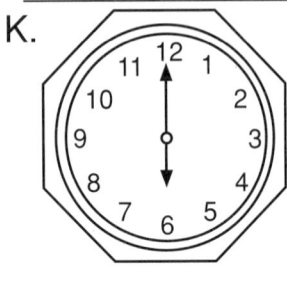

L.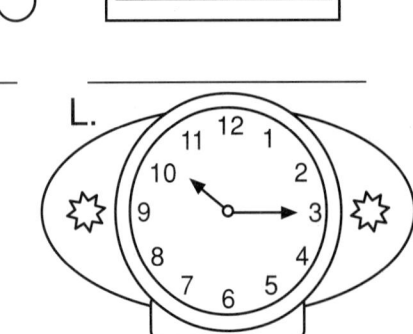

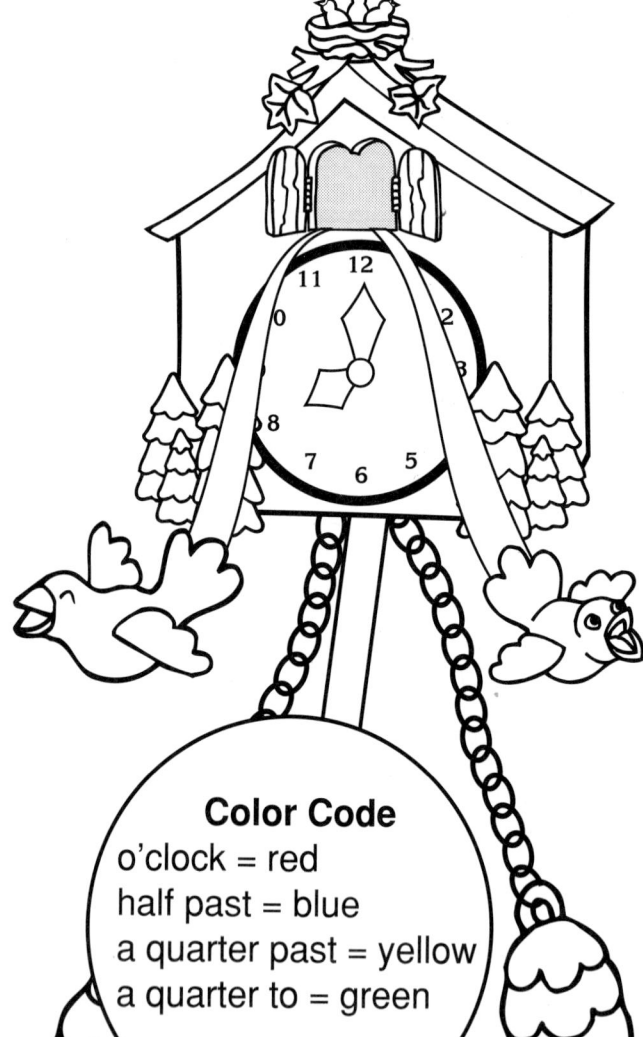

Color Code
o'clock = red
half past = blue
a quarter past = yellow
a quarter to = green

Bonus Box: Circle the clock that shows the time closest to your favorite time of day. Tell why that time is your favorite.

It's Apple-Pickin' Time!

Students will have a bushel of fun telling time to five minutes!

Purpose: To tell time to five minutes

Students will do the following:
- draw clock hands to show given times
- order times from earliest to latest

Materials for each student:
- copy of page 74
- pencil

Vocabulary to review:
- minute
- earliest
- latest

Extension activities to use after the reproducible:

- Youngsters can spin their way to time-telling success with this partner activity! Give each pair of students a copy of the clock pattern on page 164, a large paper clip, and a small paper clip. Then have students use a pencil to fashion a clock spinner with hour and minute hands as shown. To use the spinner, students spin the paper clips and read the time shown to the nearest five minutes. Direct student pairs to use the spinner, in turn, and to record each time shown. After a predetermined number of times have been recorded, write how each time is read in words.

- Share the picture book *Tuesday* by David Wiesner—on a Tuesday, if desired. As you share the book, invite students to explain each illustration. Also lead students to understand the difference between A.M. and P.M. time. Then ask students to examine the last two illustrations and predict what might happen next. Direct each student to draw his prediction and write the A.M. or P.M. time of his event. If desired, have the students line up according to their times (earliest to latest) before compiling the pages sequentially into a book titled "Next Tuesday!"

Time to five minutes

Sweet Times Bakery

Give your students a sweet treat with this telling time to the minute activity.

Purpose: To tell time to the minute

Students will do the following:
- read and write times shown on analog clocks
- read and use a schedule

Materials for each student:
- copy of page 76
- pencil

Vocabulary to review:
- analog clock
- hour hand
- minute hand
- schedule

Extension activities to use after the reproducible:
- Lead students to understand the importance of telling time and reinforce a variety of time concepts by reading aloud *Telling Time: How to Tell Time on Digital and Analog Clocks* by Jules Older. Then ask students to imagine what life would be like if there were no such things as clocks, calendars, or telling time. After students share their thoughts, guide them through a desired prewriting strategy before having each youngster write an imaginative narrative about a day without time.

- Make every minute count with this quick activity that reinforces students' time-telling skills. Program each of several index cards with a different time. Whenever a few extra moments are available, select a student to draw a card. Then have her read the time on the card and manipulate the hands of a demonstration clock to show the time. Have student volunteers state a variety of ways for reading the time shown before selecting another student to pull a card.

Time to the minute

Name _____ Time to the minute

Sweet Times Bakery

There is lots of work to be done at Sweet Times Bakery!
For each clock, write the time shown.
Read the baker's schedule.
Then, on the matching line, write what the baker will do at that time.

Baker's Schedule

7:00–8:30	Make muffins.
8:30–9:30	Make dough for sweet rolls.
9:30–10:25	Bake cookies and cakes.
10:25–12:00	Take and fill orders.
12:00–12:20	Take a lunch break.
12:20–3:00	Make fruit pies.
3:00–5:00	Clean and mop kitchen.

A.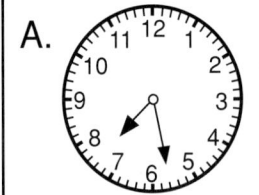

B.

C.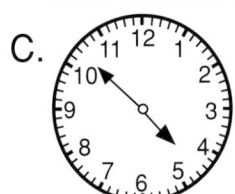

D.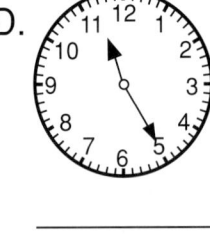

E.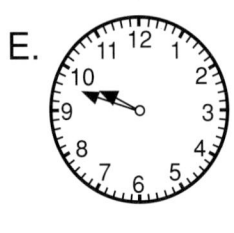

A. _____

B. _____

C. _____

D. _____

E. _____

F. _____

G. _____

H. _____

I. _____

J. _____

F.

G.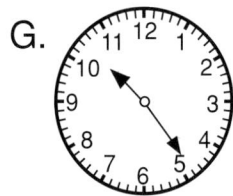

H.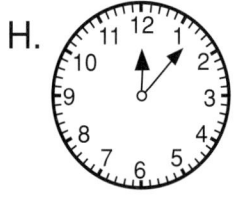

I.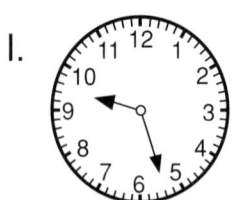

J.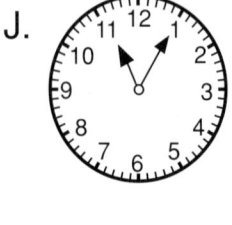

Bonus Box: Pretend you are a baker. Make up a schedule for your day.

Time Twists

Add a twist to the concept of elapsed time!

Purpose: To understand elapsed time

Students will do the following:
- solve word problems
- find elapsed time in minutes
- find elapsed time in hours and minutes

Materials for each student:
- copy of page 78
- pencil

Vocabulary to review:
- hour
- minute
- elapsed time

Extension activities to use after the reproducible:

- This real-life math activity offers plenty of practice with elapsed time. Provide each student with a construction paper copy of the clock pattern on page 164 and a brad. Instruct each student to cut out the pattern pieces and then assemble her clock. Explain to students that you are going to read aloud elapsed-time problems about your school day. For example: "At 10:10 we are in music class. Thirty-five minutes later we'll return to our classroom. At what time will we return to our classroom?" Demonstrate how each student should display her answer *(10:45)* on her clock. After each question, have a volunteer share her answer. If needed, work through the problem-solving steps as a class. What a timely way to reinforce math skills!

- Play a quick game of Time Traveler! Have two students stand facing the front of the room. Write a starting time on the chalkboard, such as 8:30. Next, hold up a card on which you've written an ending time, such as 9:15. The first player to correctly call out the elapsed time *(45 minutes)* is the winner and "travels" to the nearest seated classmate. This seated classmate stands, while the winner's first opponent sits down. The winner and the standing student then challenge each other. A game ends when every student has had a turn. The winner is the player who has won the most challenges. If desired, have the winner flash the ending time cards in the next game.

Elapsed time

Elapsed time

Time Twists

Pretzel Palace — Open 7 Days a Week

The Pretzel Guy needs your help planning his day.
Read each problem.
Solve each problem.
Write the answer.

1. The Pretzel Guy leaves his house at 7:10 A.M. He opens his shop at 7:30 A.M. How long does it take him to get to work?
   _____ minutes

2. The Pretzel Guy puts a batch of plain pretzels in the oven at 7:45 A.M. They are ready at 8:00 A.M. How long does it take to bake the batch of pretzels?
   _____ minutes

3. A batch of chocolate chip pretzels goes in the oven at 8:00 A.M. They are done at 8:40 A.M. How long does it take the batch of pretzels to bake?
   _____ minutes

4. The Pretzel Guy expects a salt delivery at 10:10 A.M. Right now it is 8:50 A.M. How long must he wait for the salt delivery truck to arrive?
   _____ hour _____ minutes

5. The Pretzel Guy mixes a batch of pretzel dough at 1:10 P.M. He runs out of dough at 2:25 P.M. How long does the dough last?
   _____ hour _____ minutes

6. A school orders a dozen chocolate chip pretzels at 12:30 P.M. The Pretzel Guy delivers the pretzels at 1:15 P.M. How long does it take to bake and then deliver the pretzels?
   _____ minutes

7. Jimmy buys a pretzel at 3:05 P.M. He finishes eating it at 3:35 P.M. How long does it take Jimmy to eat his pretzel?
   _____ minutes

8. Alice buys a pretzel at 3:15 P.M. She finishes eating it at 3:35 P.M. How long does it take Alice to eat her pretzel?
   _____ minutes

9. The Pretzel Guy puts a batch of cinnamon-raisin pretzels in the oven at 3:25 P.M. They are ready at 3:50 P.M. How long does it take to bake the batch of pretzels?
   _____ minutes

10. The Pretzel Guy closes the shop at 5:30 P.M. to clean the ovens. He goes home at 6:55 P.M. How long does it take him to clean the ovens?
    _____ hour _____ minutes

Bonus Box: How long is the Pretzel Guy at work?

©2001 The Education Center, Inc. • *Math Skills Workout* • TEC3227 • Key p. 172

Money in the Bank

Take these equivalent coin combinations all the way to the bank!

Purpose: To identify equivalent sets of coins

Students will do the following:
- use a code to draw coin sets
- find the value of coin sets
- compare coin sets

Materials for each student:
- copy of page 80
- pencil
- crayons

Vocabulary to review:
- equivalent
- combination
- compare

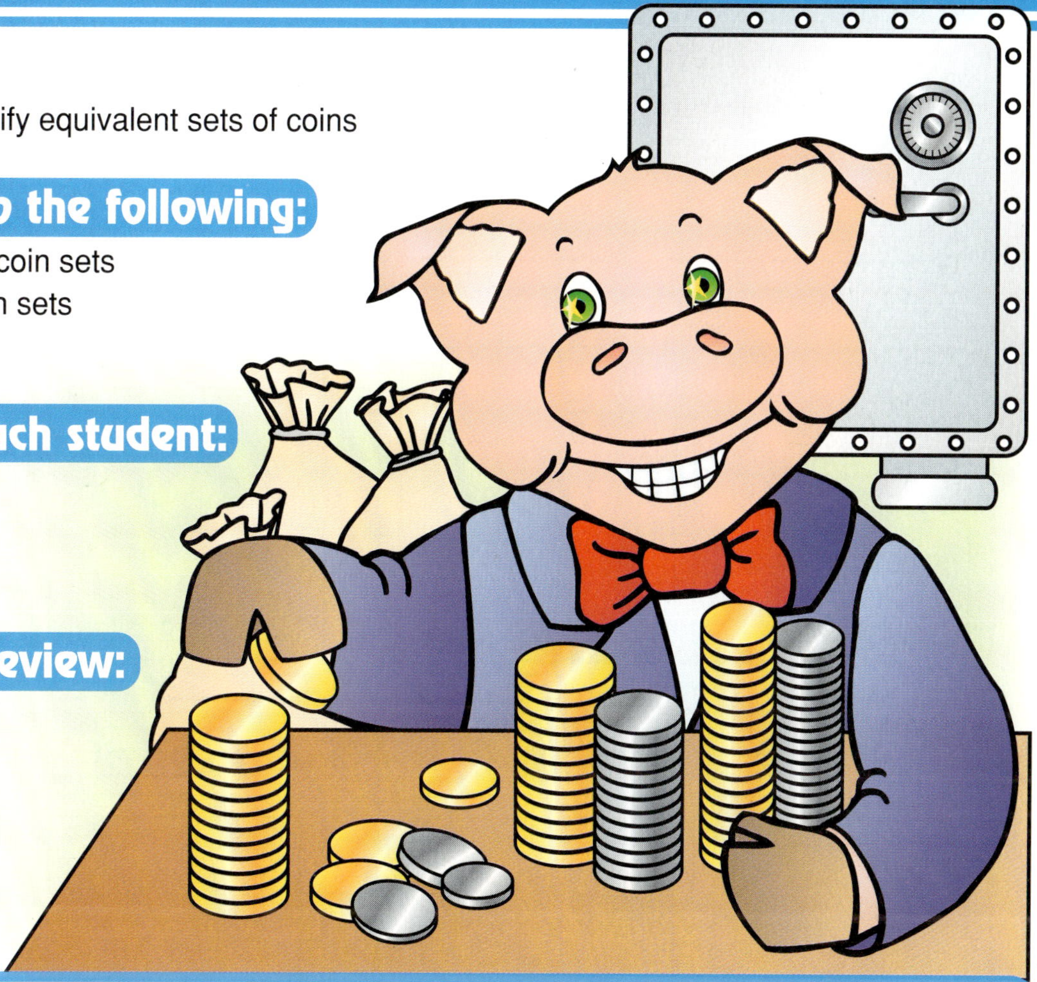

Extension activities to use after the reproducible:

- Set up a math center titled "Scrambled Coins," at which two students challenge each other to determine equivalent sets of coins. First, program the 12 cups of a Styrofoam® egg carton with coin values of 1¢, 5¢, 10¢, 25¢, and 50¢. (Use some coin values two or three times each.) Add five small chenille pom-poms to the carton and provide a set of play money coins. To play, one student shakes the closed egg carton, sets it down, and then opens it. Both students then count the total amount of money shown by the placement of the pom-poms. Next, each student uses the play money coins to make a different set of coins that is equivalent to the amount shown by the pom-poms. The first student to make a correct combination wins that round.

- Provide each student with two index cards; then assign each one a different amount of money from 25¢ to 99¢. Have the student draw two different combinations of coins—one on each card—that are equivalent to her assigned amount. (See the examples.) Check each pair of cards; then collect them to make one deck. To play Go to the Bank, have four students sit in a circle and give the deck of student-made cards to a dealer. The dealer deals five cards to each player, placing the remaining ones facedown. Students play Go to the Bank in the same fashion as Go Fish as they try to match cards that show equivalent coin combinations.

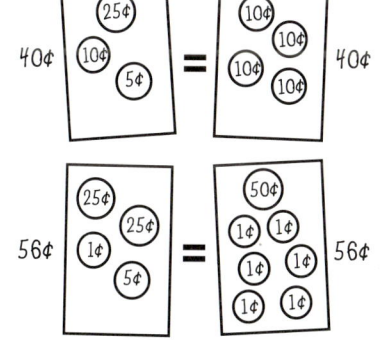

Comparing coin sets

Name _____   Comparing coin sets

Money in the Bank

Use the combination code to figure out how much money is in each safe.
On each safe, draw the coins. Then write the total amount in the blank.
In each pair, color the safe that has more money.
One safe has been completed for you.

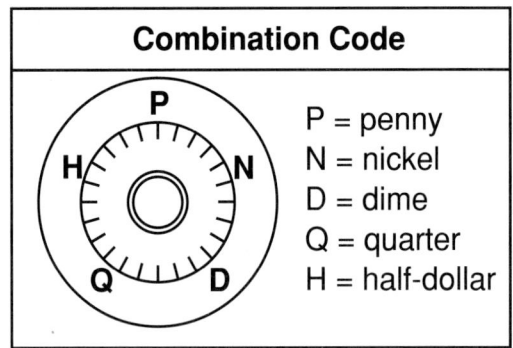

Combination Code

P = penny
N = nickel
D = dime
Q = quarter
H = half-dollar

1. 3Q, 1N, 3P | 2Q, 2D, 3P
   83 ¢   ___ ¢

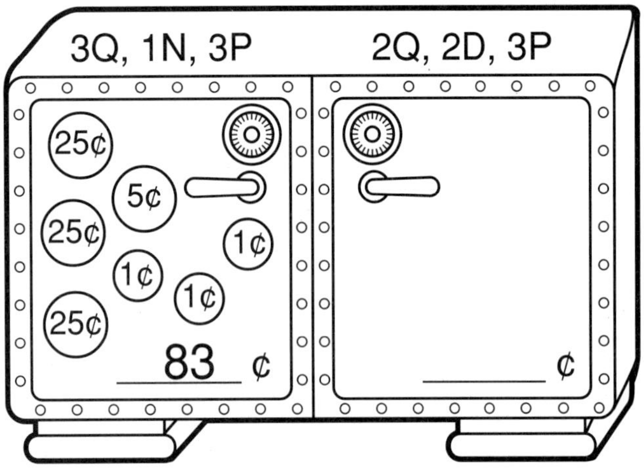

2. 3Q, 2N, 4P | 1H, 1Q, 1P
   ___ ¢   ___ ¢

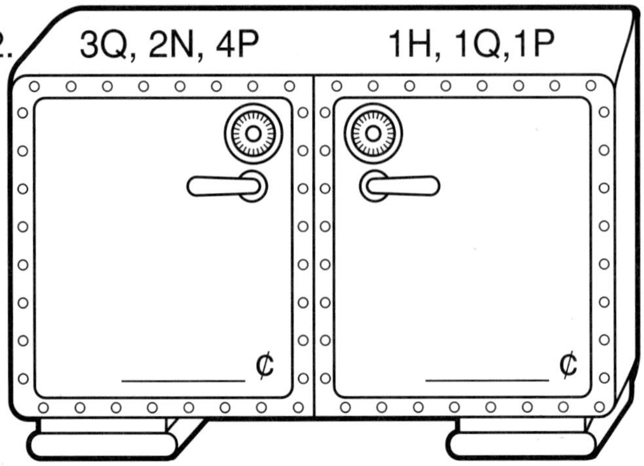

3. 1Q, 4D, 4P | 1H, 4N, 4P
   ___ ¢   ___ ¢

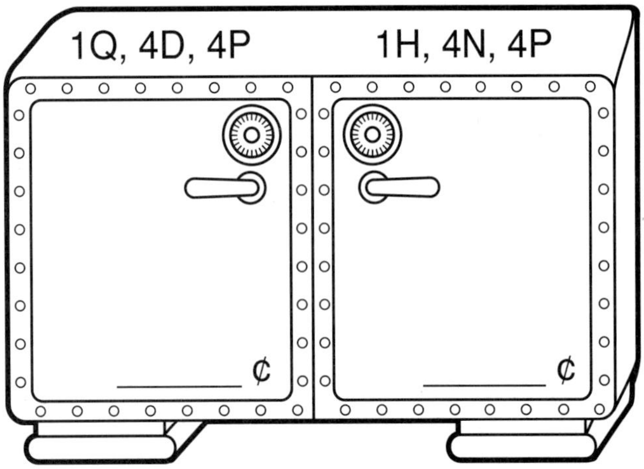

4. 3Q, 2D, 3P | 2Q, 3D, 3N
   ___ ¢   ___ ¢

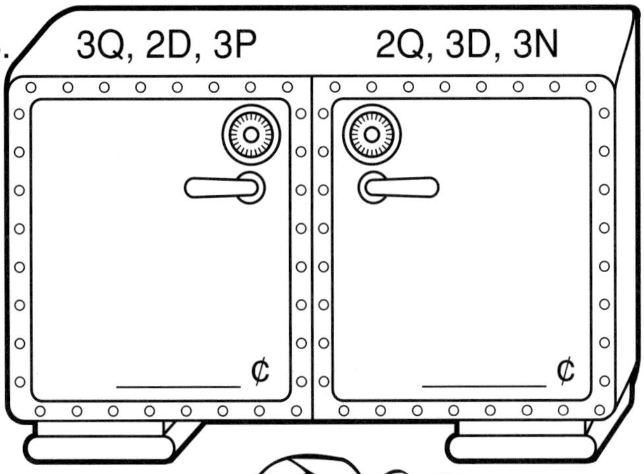

5. 3D, 2N, 2P | 1Q, 1D, 1N
   ___ ¢   ___ ¢

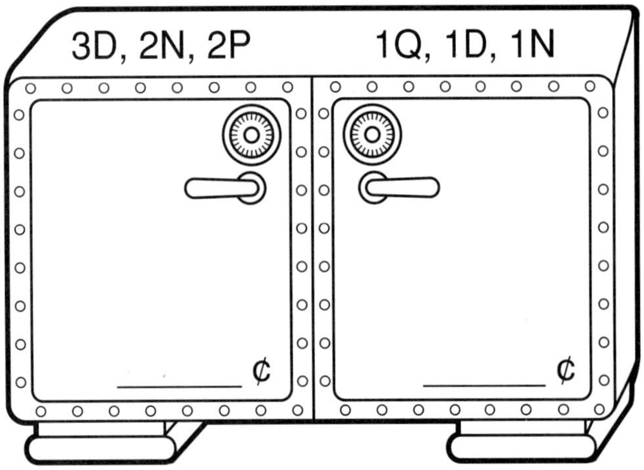

Bonus Box: Use the combination code to write three different coin combinations that equal 75¢.

80   ©2001 The Education Center, Inc. • Math Skills Workout • TEC3227 • Key p. 172

Snack Attack With Money Back

Count up to determine the change due from these tasty snack purchases!

Purpose: To determine correct change

Students will do the following:
- count a given amount of money
- determine the correct change

Materials for each student:
- copy of page 82
- pencil

Vocabulary to review:
- change
- counting up

Extension activities to use after the reproducible:

- Host a students' arts-and-crafts sale to provide a real-life opportunity for making change. Plan ahead to allow students time to make a variety of seasonal crafts. Enlist parent volunteers to help students organize and price items. Invite parents to attend and shop! Have students handle each purchase, count the money, and make change. Donate the proceeds from the sale to a charity of your students' choice.

- Finding the correct change will be in the bag with this one-of-a-kind idea! Give each student a small paper bag and an index card. Have each student follow the directions below to make a page for a "Money Bags" activity book. Collect the completed bags. For every five bags, punch two holes in the bottom of each bag. Then decorate and hole-punch a cover bag in the same manner, place it on top, and combine the bags with yarn to make a booklet as shown. Next, place a variety of play money coins and bills in a resealable plastic bag and tuck it inside the cover bag. Place the booklets in your math center. A student reads each problem in a booklet, uses the play money in the cover to solve each problem, and then looks at the index cards inside the bags to check his answers.

Directions:
1. Turn the bag so that the open end is to your right and the closed end is to your left.
2. Cut out and glue or draw a picture of an item worth $3.00 or less on your bag.
3. Write a story problem about the picture. Include the cost of the item, the amount of money given to pay for the item, and a question about change.
4. Write how to solve the problem and the correct answer on the index card and place the card inside the bag.

Making change

Snack Attack With Money Back

Making change

Complete the chart below.
Write the cost of the snack. Look at how much money you have.
Then count up to find how much change you should get.
Write the number of each coin you should receive.
The first one is done for you.

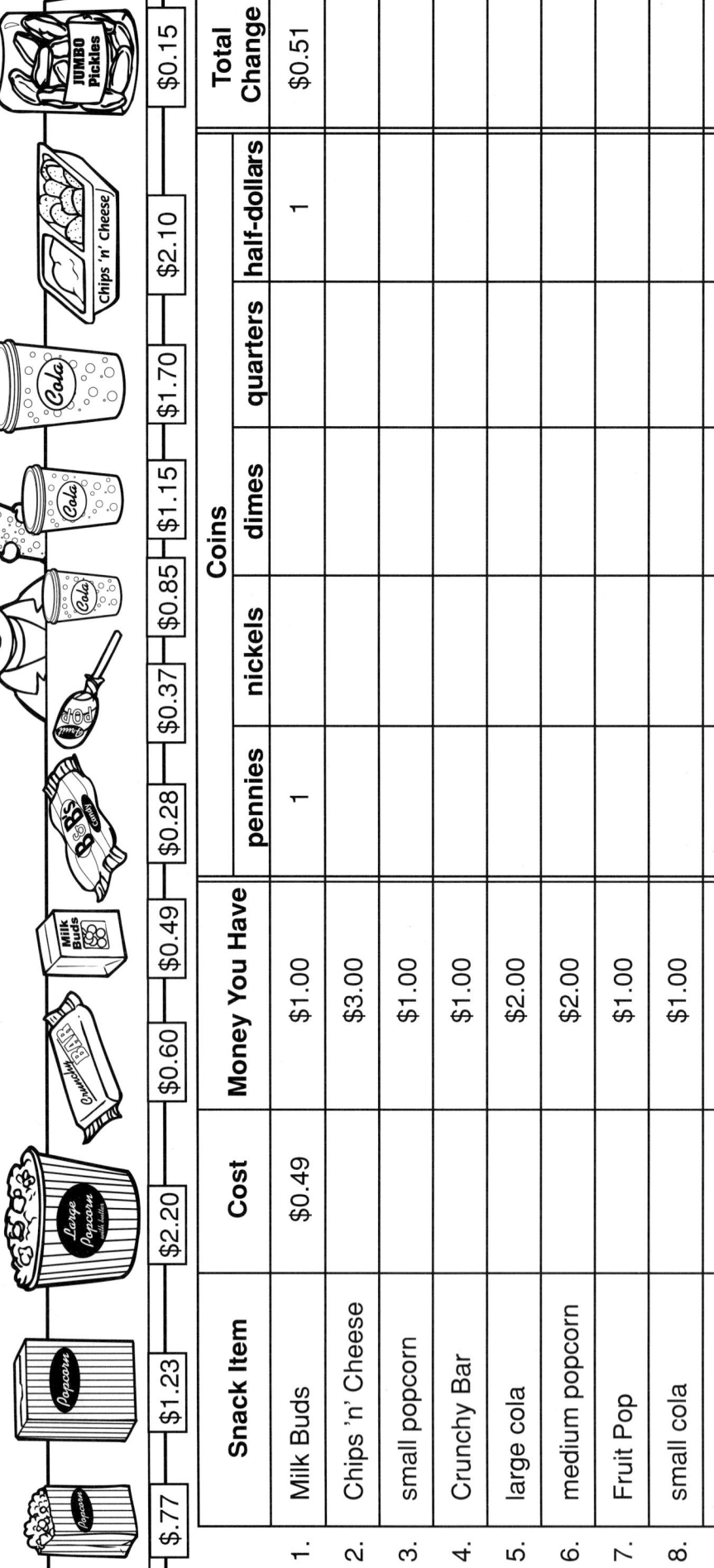

Main Street MOVIE SNACKS

	Snack Item	Cost	Money You Have	Coins					Total Change
				pennies	nickels	dimes	quarters	half-dollars	
1.	Milk Buds	$0.49	$1.00	1				1	$0.51
2.	Chips 'n' Cheese		$3.00						
3.	small popcorn		$1.00						
4.	Crunchy Bar		$1.00						
5.	large cola		$2.00						
6.	medium popcorn		$2.00						
7.	Fruit Pop		$1.00						
8.	small cola		$1.00						
9.	B & B's		$1.00						
10.	large popcorn		$3.00						

Get Set to Own a Pet

Buy a pet, but don't forget that the cost adds up!

Purpose: To add and subtract money

Students will do the following:
- choose appropriate information to solve problems
- select addition or subtraction to solve problems
- add and subtract money

Materials for each student:
- copy of page 84
- pencil

Vocabulary to review:
- money
- change

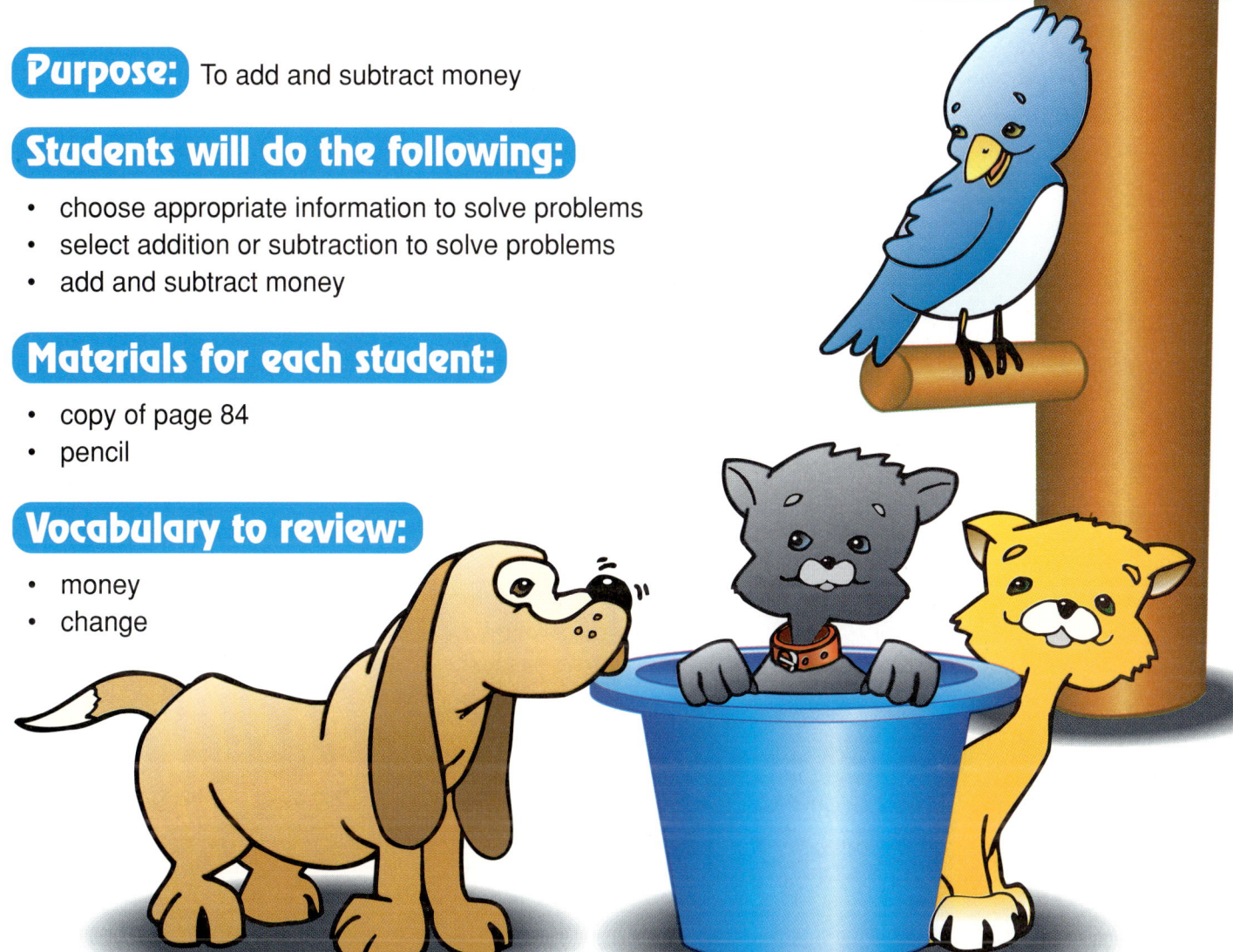

Extension activities to use after the reproducible:
- Liven up the skills of adding and subtracting money with this motivating activity. First, devise a plan in which students can earn "money." (For example, a student may earn 25¢ for completing an assignment, 30¢ for helping a classmate, and 40¢ for tidying up the classroom library.) Have each student keep a running total of his earnings for a predetermined time period. Then plan an ice-cream party at which students can spend their earned money. Assign a price to each item offered. (For example, a scoop of ice cream costs 50¢, chocolate topping costs 25¢, and candy sprinkles cost 15¢.) Have each student determine the total amount of his ice-cream order. Then have him subtract that total from his earnings to find how much money he has left over. Math never tasted this good!
- Collect a variety of catalogs for students to use for adding and subtracting money. Assign a specific amount that each student may spend. Then challenge students to shop till they drop! Have students add to find out the total of their purchases and then subtract that total from the amount with which they began.

Adding and subtracting money

Get Set to Own a Pet

Pete's Pet Palace sells lots of pets and all the supplies needed to care for them. Use the information on the signs to solve each problem. Show your work.

Cats & Kittens
- cat $6.00
- kitten $5.00
- dry kitten food $3.50
- canned kitten food .. $0.99
- kitten toy $1.50
- cat bed $5.49

Puppies
- puppy $8.00
- dry puppy food $3.00
- canned puppy food .. $2.50
- chew bone $0.75
- leash $1.25
- collar $2.00

Parakeets
- parakeet $3.00
- cage $6.50
- birdseed $2.25
- salt block $1.75
- cage toy $1.75

Fish
- goldfish $1.25
- fishbowl $2.50
- fish food $2.00
- gravel $1.75
- seaweed $0.98
- snail $0.50

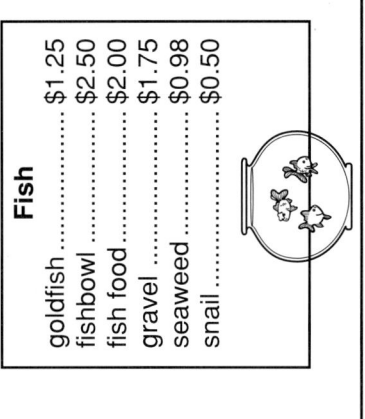

1. How much would a kitten, dry kitten food, and a kitten toy cost?

2. If you give the clerk $6.00 to buy a cat bed, how much change will you receive?

3. How much would a puppy, canned puppy food, and a collar cost?

4. How much would a chew bone and a leash cost for your new puppy?

5. If you give the clerk $10.00 for a puppy, how much change will you receive?

6. How much is a parakeet, cage, and birdseed?

7. How much is a cage toy and a salt block for your new parakeet?

8. What does it cost to buy two goldfish, a fishbowl, and fish food?

9. If you give the clerk $10.00 for the items in problem 8, how much change will you receive?

10. How much will you spend to buy gravel and seaweed for the fishbowl?

Bonus Box: Look at your answers for problems 1, 3, 6, and 8. Which pet (with basic supplies) costs the most? Which costs the least?

Juggling Polygons

Students will have a ball with this shapely activity!

Purpose: To identify and construct polygons

Students will do the following:
- solve riddles about polygons
- construct polygons

Materials for each student:
- copy of page 86
- pencil

Vocabulary to review:
- polygons: triangle, square, rectangle, pentagon, hexagon, octagon, trapezoid, parallelogram
- parallel lines
- right angles

Extension activities to use after the reproducible:

- Stretch students' polygon knowledge with this Geoboard activity! Program a spinner as shown (pattern on page 166). Then pair students and give each pair two Geoboards and two large rubber bands. Invite a student volunteer to spin the spinner. Instruct each twosome to display two different ways to create the selected polygon. Then have students share a range of possible answers. Repeat this process for as long as time allows, having a new volunteer spin each time.

- Get students' understanding of polygons in shape by reading aloud *The Greedy Triangle* by Marilyn Burns. Before you read the story, give each group of four students a five-foot length of yarn with its ends tied together. Tell students that they will play the role of the shapeshifter in the story. As you read the story aloud, pause each time the shapeshifter changes the triangle and have each group manipulate its yarn to form the new polygon.

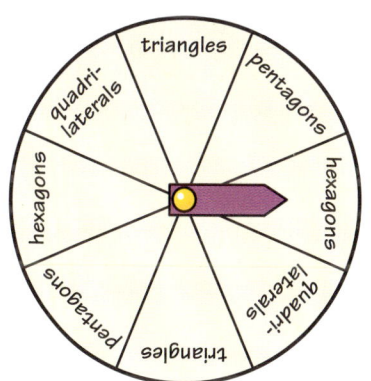

Polygons 85

Name _____   Polygons

Juggling Polygons

Read the clue on each ball.
Draw and label the correct polygon in the matching box.
Use the word bank to help you.

Word Bank
pentagon   square   triangle
rectangle   hexagon   trapezoid
octagon   parallelogram

1. I have twice as many sides as a square.
2. I am a 3-sided polygon.
3. I have 4 equal sides and 4 right angles.
4. The number of my sides is half of 10.
5. I have 4 right angles, but I am not a square.
6. I have twice as many sides as a triangle.
7. I have 4 sides. My opposite sides are the same length and parallel.
8. I have 4 sides with only 1 pair of parallel lines.

1.	2.	3.	4.
5.	6.	7.	8.

Bonus Box: A *quadrilateral* is a polygon with 4 sides. Draw 4 different quadrilaterals.

86   ©2001 The Education Center, Inc. • *Math Skills Workout* • TEC3227 • Key p. 172

Polygon Parade

March along as students identify and classify polygons!

Purpose: To identify and classify polygons

Students will do the following:
- identify pictures of polygons
- sort polygons

Materials for each student:
- copy of page 88
- pencil
- crayons
- scissors
- glue

Vocabulary to review:
- polygons: triangle, quadrilateral, pentagon, hexagon

Extension activities to use after the reproducible:
- Give your youngsters a new angle on polygons with tangrams! Create and post a desired number of tangram challenges similar to those shown. Pair students and provide each pair with a tangram set and drawing paper. Have each pair complete each challenge and illustrate its answer by drawing or tracing the tangrams.

- Add a picturesque flair to polygons with this pattern block activity! Give each student a set of pattern blocks and have him arrange the blocks—on the top half of a sheet of drawing paper—into an imaginary pet. Direct each student to use a pencil to trace his pet's outline and then color it. Next, instruct each student to write on the bottom half of his paper about how his pet was constructed, including accurate names for the polygons used. If desired, combine students' work into a book titled "Polygon Pets" and place it in your classroom library for all to enjoy.

Tangram Challenges
- Make a parallelogram using two pieces.
- Make a pentagon using three pieces.
- Make a triangle using four pieces.
- Make a trapezoid using six pieces.
- Make a square using seven pieces.

Polygons

Name _____   Polygons

Polygon Parade

Use the code to color the shapes below. Then cut out each polygon and glue it in the correct parade line.

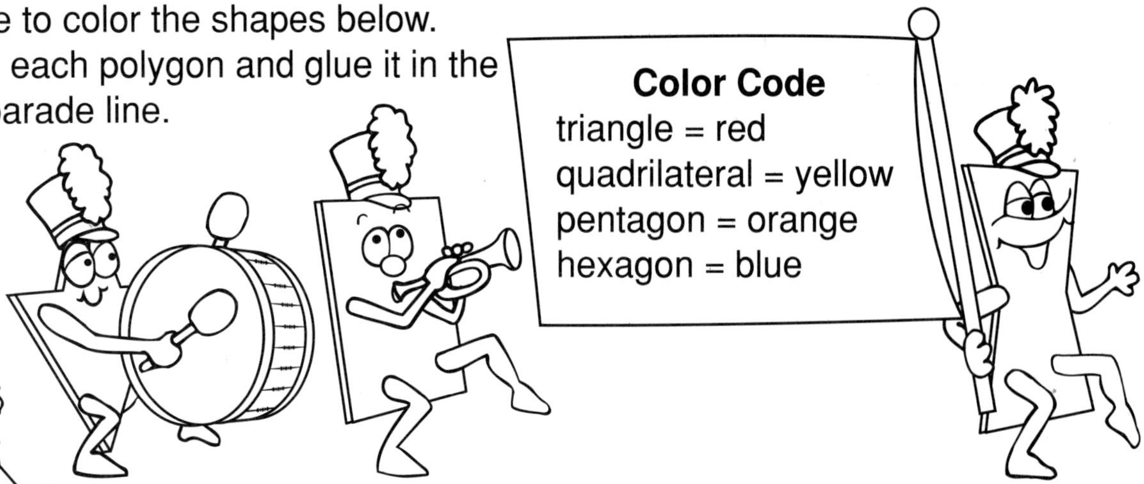

Color Code
triangle = red
quadrilateral = yellow
pentagon = orange
hexagon = blue

Triangle Parade

Quadrilateral Parade

Pentagon Parade

Hexagon Parade

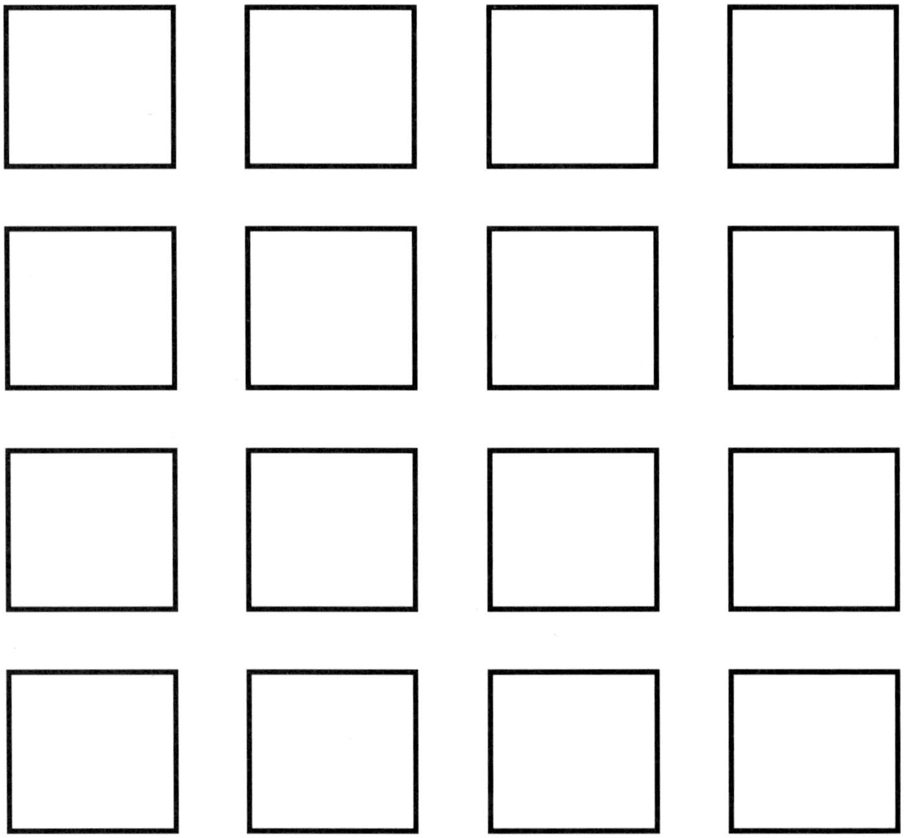

Bonus Box: Write the name of each polygon in the quadrilateral parade.

©2001 The Education Center, Inc. • Math Skills Workout • TEC3227 • Key p. 173

Shopping for Space Figures

Send your youngsters on a space figures shopping spree!

Purpose: To identify space figures

Students will do the following:
- identify space figures
- label real-life examples of space figures

Materials for each student:
- copy of page 90
- pencil

Vocabulary to review:
- space figures
- sphere
- cube
- cylinder
- pyramid
- rectangular prism
- cone

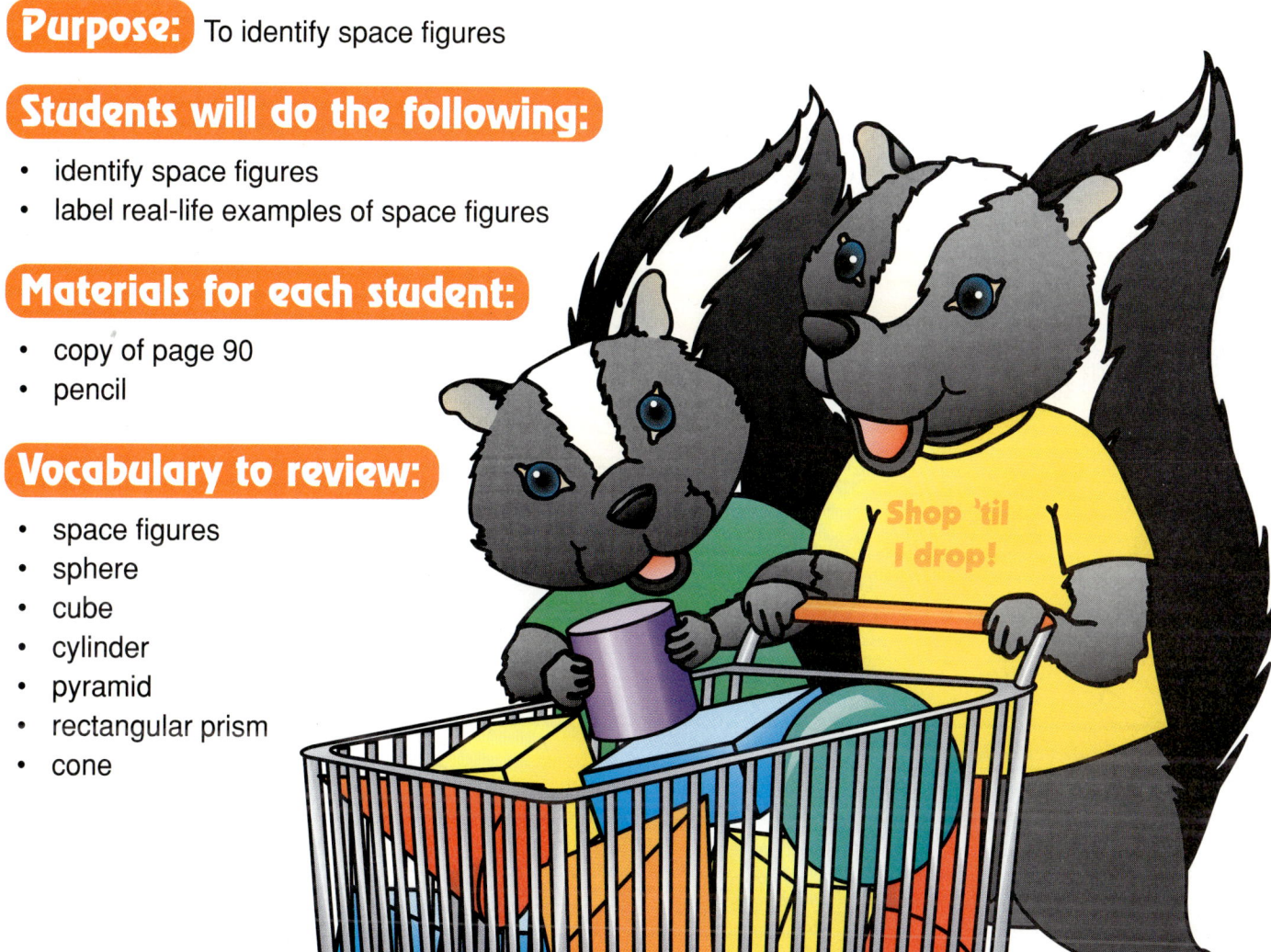

Extension activities to use after the reproducible:
- Use this sweet activity to give students practice classifying space figures and graphing data. Each small group of students will need a bag of assorted wrapped candies that represent different space figure shapes (for example, Tootsie Roll® candy for cylinders, caramels for cubes, or jawbreakers for spheres). Instruct groups to organize and discuss the contents of their bags. Then have each group member use a copy of the one-inch graph paper on page 168 to independently create a bar graph representing her group's data. Invite a volunteer from each group to share her graph and discuss her group's findings. If desired, give each student one piece of candy and then collect the rest to use later as special classroom treats.

- Test your youngsters' knowledge of space figures with this riddle review! In advance, gather a class supply of large index cards. Randomly program the back of each card with one of the following space figures: rectangular prism, cone, cube, cylinder, pyramid, or sphere; then distribute one card to each student. Instruct each student to write a riddle on the front of her card about her assigned space figure. Provide time for students to share their riddles with classmates.

Space figures

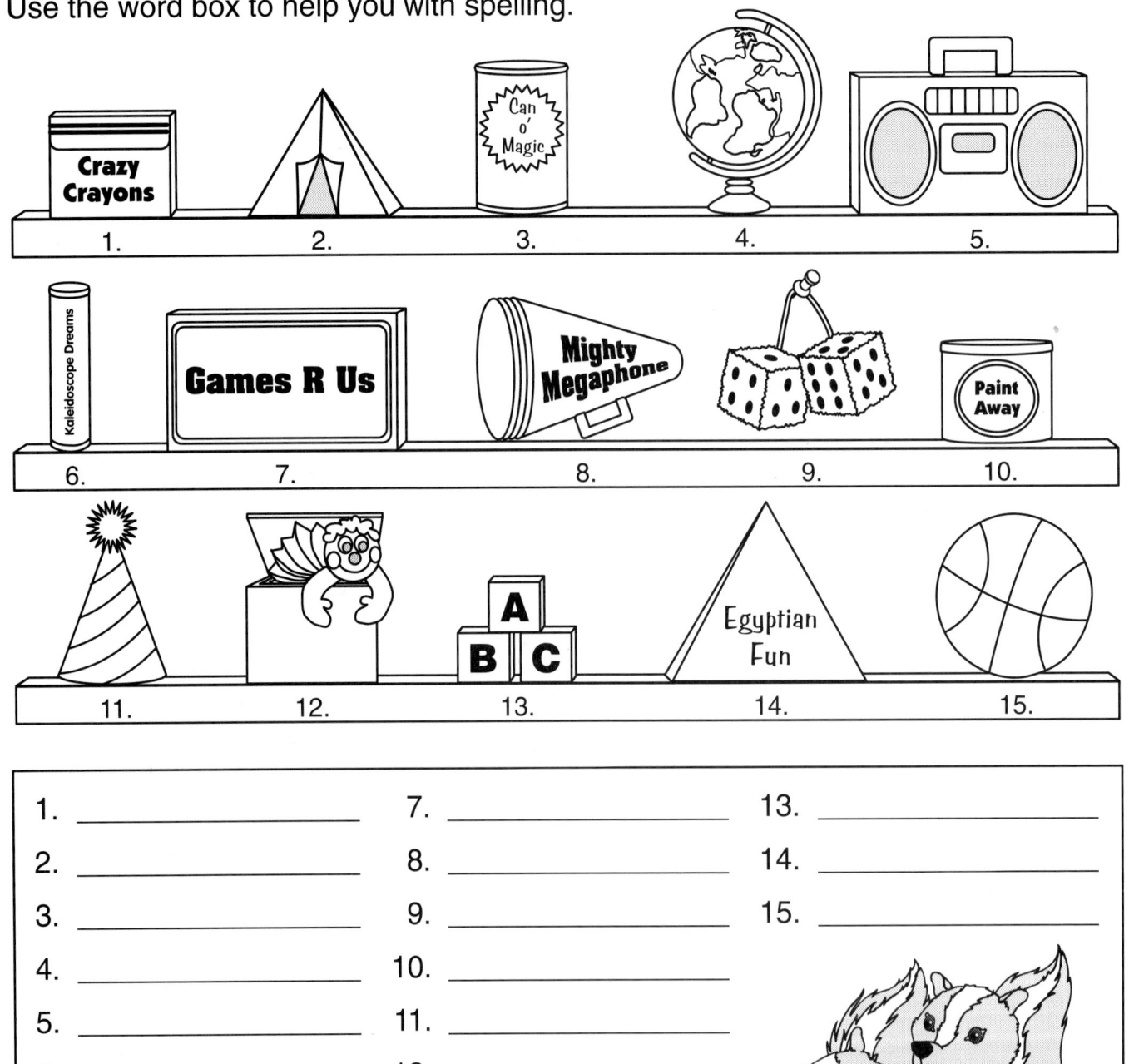

Exploring Space Figures

Add a new dimension to students' understanding of space figures!

Purpose: To identify attributes of space figures

Students will do the following:
- identify and count the number of faces, edges, and corners of space figures
- label space figures and space figure combinations according to attributes
- complete a chart

Materials for each student:
- copy of page 92
- pencil

Vocabulary to review:
- space figures: cone, cube, cylinder, pyramid, rectangular prism, sphere
- faces
- edges
- corners

Extension activities to use after the reproducible:
- Extend the activity on page 92 to give youngsters practice comparing the attributes of space figures. Draw a large Venn diagram on the chalkboard. Then invite a student volunteer to roll a pair of dice. (If doubles are rolled, have the student continue to roll until two different numbers appear.) Direct students to identify the two space figures on their copies of page 92 that correspond with the two numbers shown on the dice. Label each section of the Venn diagram accordingly. Then, as students brainstorm how the two space figures are alike and different, list their ideas on the diagram. Repeat this process for as long as time allows.

- Provide more practice with space figures by having each student create a personalized picture dictionary. Each student will need seven sheets of drawing paper and access to a supply of discarded magazines. To make a dictionary, each student illustrates a cover for the picture dictionary on one sheet of paper. Then she writes one of the following terms at the top of each remaining sheet: *cone, cube, cylinder, pyramid, rectangular prism,* and *sphere*. Next, she looks through old magazines to find an example of each space figure, cuts it out, and pastes it on the appropriate sheet of paper. Finally, the student writes a definition for each space figure that includes the number of faces, edges, and corners the space figure has. Then she stacks the completed sheets in alphabetical order underneath the cover and staples along the left side to create a picture dictionary booklet.

Space figures

Name _____   Space figures

Exploring Space Figures

Mission #1: Study each spacecraft. In the matching space on the chart, write the shape of the spacecraft. Use the word bank for help with spelling. Then complete the chart.

1.

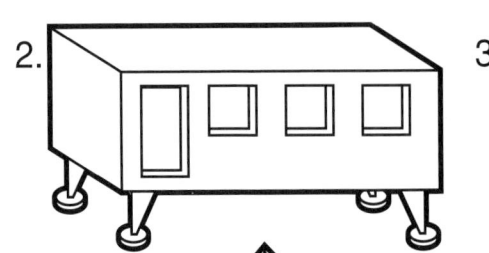
2.

3.

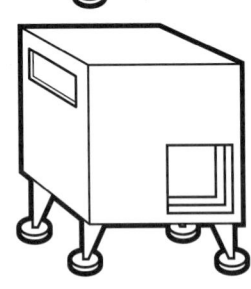
4.

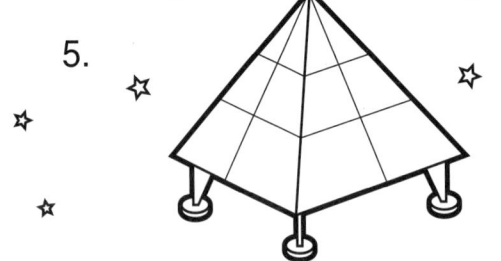
5.

6.

Word Bank
- sphere
- cone
- pyramid
- cylinder
- cube
- rectangular prism

Spacecraft Shape	Number of Faces	Number of Edges	Number of Corners
1.			
2.			
3.			
4.			
5.			
6.			

Mission #2: Each spacecraft is made up of 2 space figures. List the 2 space figures.

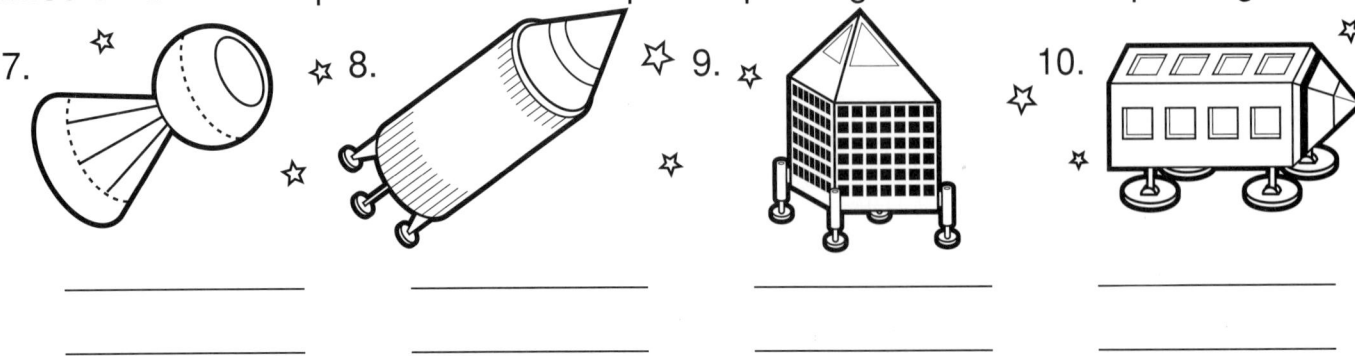

7. _____   8. _____   9. _____   10. _____

Bonus Box: Write an adventure about 1 of the spacecrafts shown above. Describe the spacecraft in your story.

Dot-to-Dot and Beyond!

Help students make an out-of-this-world connection between lines, line segments, and rays!

Purpose: To identify and construct lines, line segments, and rays

Students will do the following:
- identify lines, line segments, and rays
- construct lines, line segments, and rays
- create a picture using lines, line segments, and rays

Materials for each pair of students:
- copy of page 94
- pencil
- ruler

Vocabulary to review:
- line
- line segment
- ray

Extension activities to use after the reproducible:

- This drawing activity is just perfect for reviewing lines, line segments, and rays, as well as reinforcing skills in listening and following directions! Draw a simple figure and write directions for drawing the figure on a transparency as shown. Then give each student a ruler and a sheet of drawing paper. Without revealing the figure to your students, instruct them to draw the figure as you read one direction at a time. After students have completed their drawings, turn on the overhead projector so they can compare their pictures to yours.

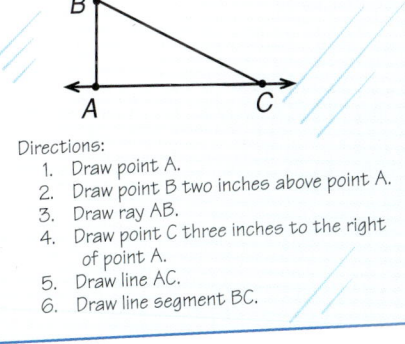

Directions:
1. Draw point A.
2. Draw point B two inches above point A.
3. Draw ray AB.
4. Draw point C three inches to the right of point A.
5. Draw line AC.
6. Draw line segment BC.

- Assess students' abilities to identify and construct lines, line segments, and rays with this personalized task. Draw on the chalkboard the geometric figure shown. Point out to students that your figure consists of lines, line segments, and rays. Give each student a sheet of drawing paper and a ruler. Then challenge each student to create a similar figure of his own within a predetermined amount of time. Next, have each student write a description of his figure that includes names of specific lines, line segments, or rays.

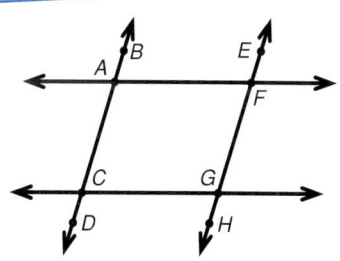

Lines, line segments, and rays

Name _____   Lines, line segments, and rays

Dot-to-Dot and Beyond!

Write the name for each figure.
The first one is done for you.

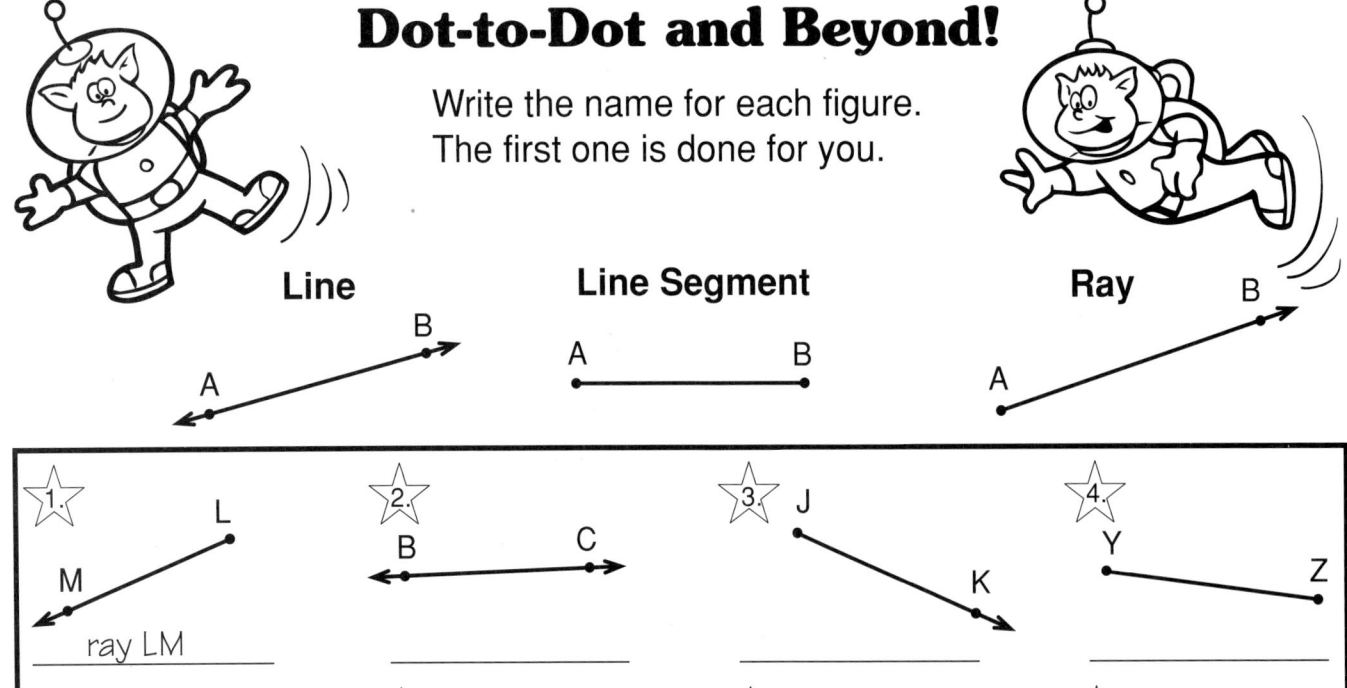

1. ray LM
2. _____
3. _____
4. _____
5. _____
6. _____
7. _____
8. _____

Make an out-of-this-world picture appear!
Use the dots and a ruler to draw each figure listed below.
Then check off each figure as you draw it.
The first one is done for you.

- ✔ line AB
- ☐ line BC
- ☐ line CD
- ☐ line DA
- ☐ line segment EF
- ☐ line segment FG
- ☐ line segment GH
- ☐ line segment HI
- ☐ line segment IJ
- ☐ line segment JK
- ☐ line segment KL
- ☐ line segment LM
- ☐ line segment MN
- ☐ line segment NE
- ☐ line segment FJ
- ☐ ray NO
- ☐ ray MP
- ☐ ray LQ

Bonus Box: Draw your own out-of-this-world picture that is made up of lines, line segments, and rays.

Awesome Origami

Watch students' symmetry skills unfold as they explore origami.

Purpose: To identify and construct symmetrical figures

Students will do the following:
- identify symmetrical figures
- identify lines of symmetry
- use paper folding to explore symmetry

Materials for each student:
- copy of page 96
- two 8" x 8" paper squares
- pencil

Vocabulary to review:
- symmetrical
- line of symmetry

Extension activities to use after the reproducible:
- Watch students' understanding of symmetry grow and grow with this interactive bulletin board display! Mount three large tree cutouts on a bulletin board titled "Growing 'Symme-trees.'" Label each tree with one of the following: *one line of symmetry, two lines of symmetry,* and *more than two lines of symmetry*. Make magazines, scissors, glue, markers, clear tape, and a supply of leaf cutouts available to students. Then, during free time or center time, direct students to search the magazines and cut out one picture for each tree. Then have each student draw the correct lines of symmetry on each picture, glue each picture to a leaf, and tape each leaf on the limb of the corresponding tree.

- Take youngsters outdoors to explore symmetry in nature. Divide students into pairs and give each pair a small plastic bag. Instruct each pair to gather two items that are symmetrical and two that are asymmetrical. Students will soon discover that items such as leaves, pinecones, seeds, and many other items found in nature are sometimes symmetrical; but items like twigs, bark, and rocks often are not symmetrical. Upon returning to the classroom, direct the students in each pair to descriptively write about their findings. Then provide time for each pair to share its findings and written descriptions.

Name _____ Symmetry

Awesome Origami

Origami is the ancient Japanese art of paper folding. Steps for creating 2 origami projects are shown below. Study the outlined shape for each step.
 If it has no lines of symmetry, write "0" on the line.
 If it has 1 line of symmetry, write "1" on the line.
 If it has more than 1 line of symmetry, write "1+" on the line.
Be careful not to let the fold lines and the arrows trick you!

How to Make the Cup

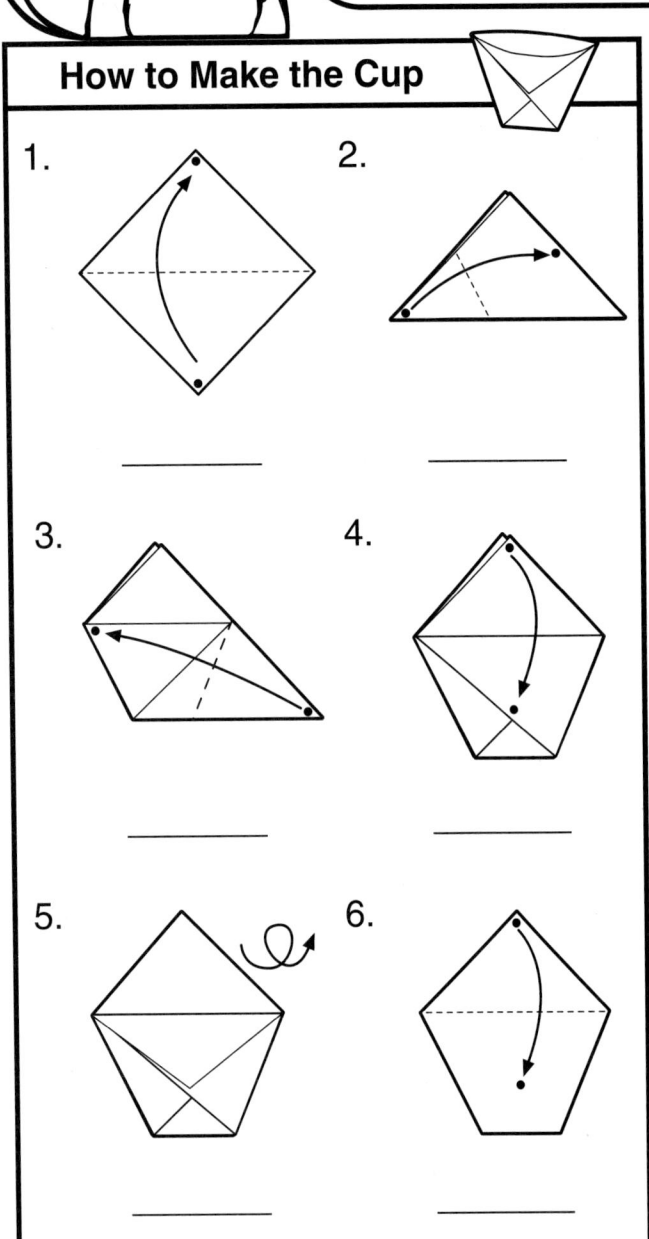

How to Make the Whale

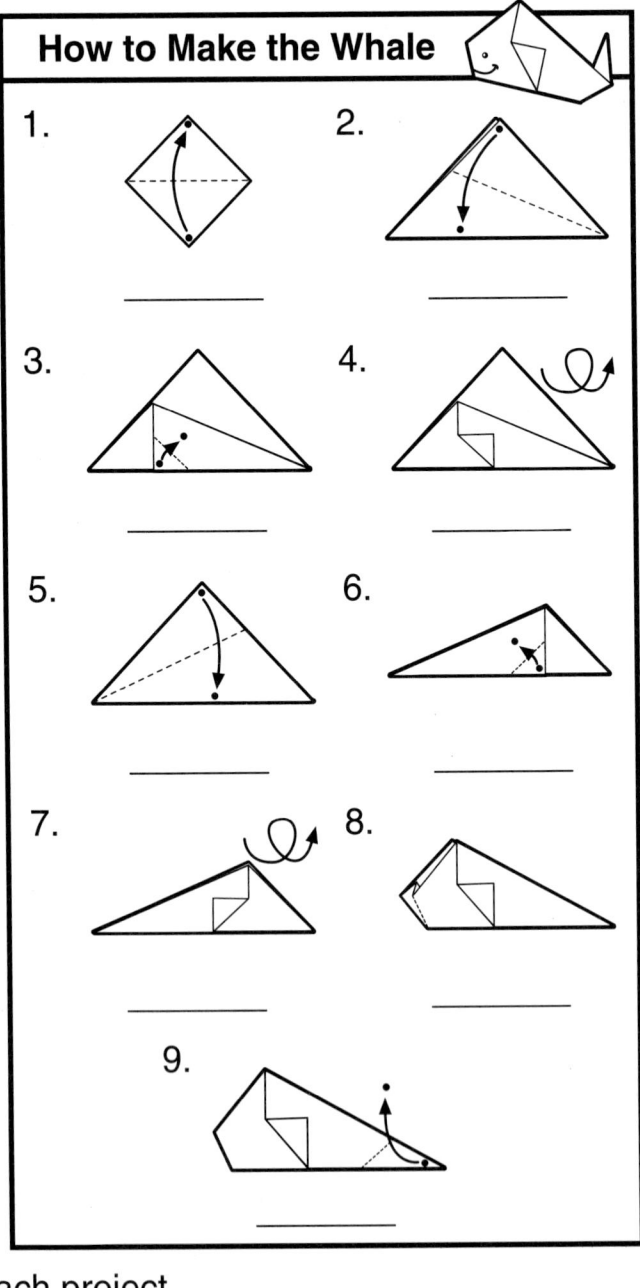

Now use your paper squares to complete each project.
Use the symbols below to help you.

- - - - - - - = fold line   = fold from dot to dot   = turn project over

96   ©2001 The Education Center, Inc. • Math Skills Workout • TEC3227 • Key p. 173

Brushing Up on Congruent Figures

Have your youngsters turn congruent figures into works of art!

Purpose: To identify congruent figures

Students will do the following:
- identify congruent figures
- draw a congruent example of a figure

Materials for each student:
- copy of page 98
- pencil
- crayons
- centimeter ruler

Vocabulary to review:
- congruent

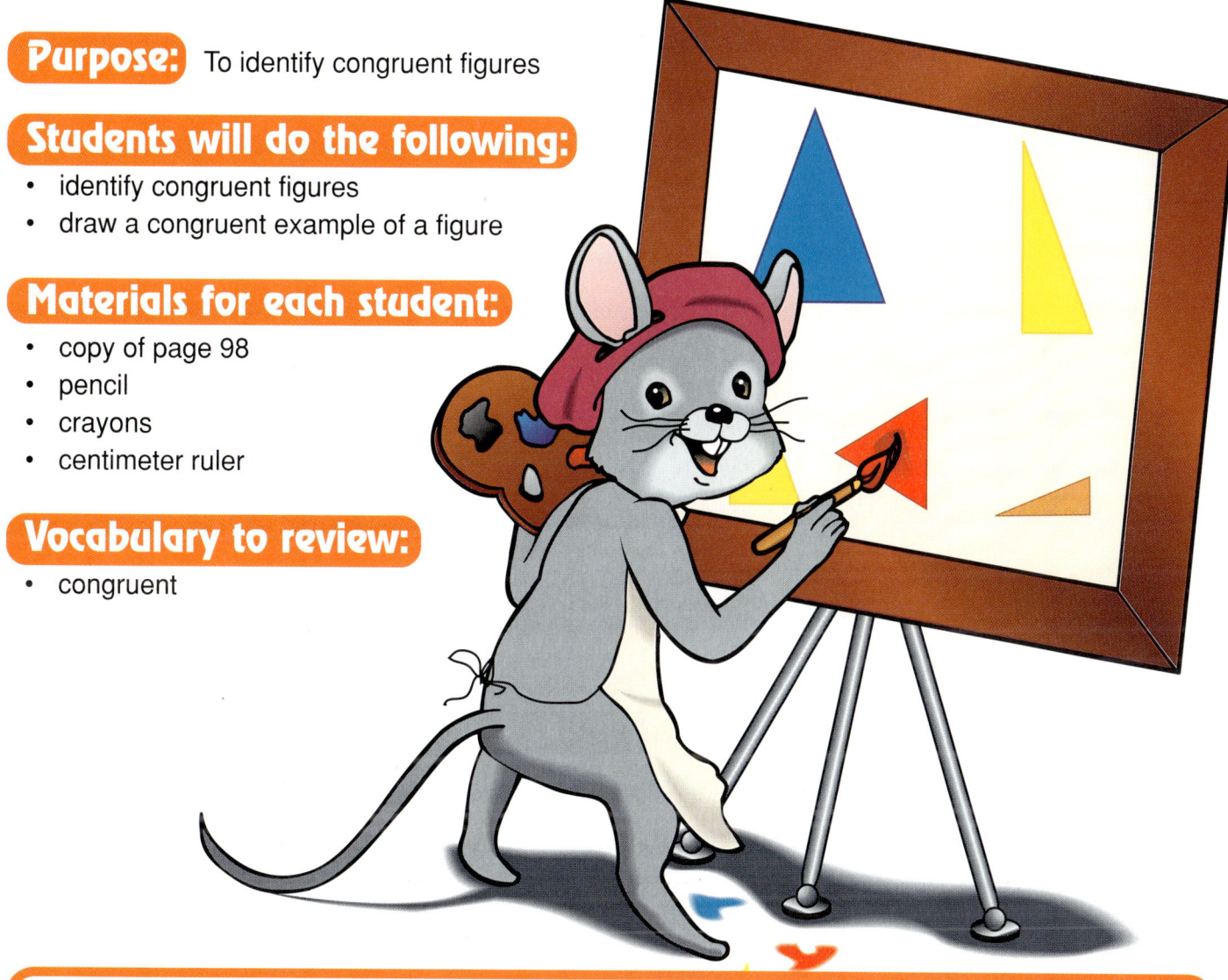

Extension activities to use after the reproducible:

- Shed light on congruent figures with this technique. Divide students into groups of four and supply each student with a set of pattern blocks. Use pattern blocks to form a unique shape on an overhead projector. Then turn on the projector so that students will see the outline of the shape. Challenge each student in the group to use pattern blocks to form the shape shown. Then have group members compare their shapes to see if they are congruent (the same size and shape) or similar (same shape but different size). Have one person from each group report the group's findings. Then repeat this process by forming a different shape on the overhead projector.

- Double the fun of finding congruent figures with this partner activity. Challenge students to determine if the left and right sides of their own bodies are actually congruent. Have students work with their partners to carefully trace their left and right hands and feet. Then have each student cut out the tracings and compare each for congruency. Next, give each pair a measuring tape. Instruct partners to help each other measure and record the lengths and circumferences of their arms and legs in centimeters to compare for congruency. Then have each student write about his findings.

Congruent figures

Angle Antics

Use these acrobatic antics to put a new angle on measurement!

Purpose: To identify and categorize angles

Students will do the following:
- identify acute, obtuse, and right angles
- sort angles into appropriate categories

Materials for each student:
- copy of page 100
- pencil
- scissors
- glue

Vocabulary to review:
- acute angle
- obtuse angle
- right angle

Extension activities to use after the reproducible:
- Students' ability to describe and construct angles will be acute after this partner activity! Pair students and provide each pair with a collection of ten polygon-shaped cutouts in a paper bag, a Geoboard, and a rubber band. In turn, one student in the pair removes a shape from the bag without revealing it to his partner and describes the shape—including its number and types of angles. He continues to describe the shape until his partner is able to use the rubber band to form the shape on the Geoboard. Students alternate roles in this fashion for a predetermined amount of time.

- Invite your students to scout for angles around the classroom. Provide each student with a sheet of drawing paper, two 1" x 3" strips of tagboard, and a brad. Instruct each student to fold her paper into three sections and write one of the following headings for each section: acute angles, right angles, obtuse angles. Then direct each student to make an angle by connecting her tagboard strips with the brad as shown. Challenge students to find and copy two examples of each type of angle. To copy the angle, have each student open or close her angle so that the inside matches the angle being measured. Next, each student carefully traces the angle in the appropriate section of her drawing paper and writes to tell where the angle was found.

Angles

Angle Antics

Read what each squad captain has to say about angles.
Then study the bold angle on each stick figure below.
Cut out each figure and glue it in the correct squad.

An **acute angle** is less than 90°.

A **right angle** is exactly 90°.

An **obtuse angle** is greater than 90°.

Acute Angle Squad

Right Angle Squad

Obtuse Angle Squad

Bonus Box: List times when the hands on a clock form a right angle, an acute angle, and an obtuse angle.

©2001 The Education Center, Inc. • Math Skills Workout • TEC3227 • Key p. 173

Geometric Groove

Get your students into the groove of geometric moves—slides, flips, and turns!

Purpose: To identify and perform slides, flips, and turns

Students will do the following:
- identify the movement of a geometric figure as a slide, a flip, or a turn

Materials for each student:
- copy of page 102
- pencil

Vocabulary to review:
- slide
- flip
- turn

Extension activities to use after the reproducible:
- Here's an activity that your students will flip for! Program each of three index cards with one of the following words: "slide," "flip," or "turn." Seat the students in a circle on the floor. Ask one student volunteer to lie in the middle of the circle in any position. Call on a second student to select a card and read it silently. Then, based on the card, the student demonstrates a slide, flip, or turn of the first student's body. Choose a student volunteer to identify the move. If his answer is correct, he chooses the next position in the center of the circle while another student selects a card. Continue to play in this manner until each student has had the opportunity to participate.

- Move your students to understand slides, flips, and turns with this patterning exercise. Provide each student with a large sheet of white construction paper and a stencil or cutout of the first letter of his first or last name. Instruct each student to determine a pattern rule (such as *slide, flip, flip* or *turn, slide, flip*) and write it on the back of his paper. Then direct each student to place his stencil or cutout on the front of his paper and trace it to show his pattern. Have students color their completed patterns. Provide time for each student to present his illustrated pattern and challenge classmates to name his pattern rule.

Slides, flips, and turns

Name_____   *Slides, flips, and turns*

Geometric Groove

Dancing Danny knows how to groove!
Study the examples.
Then identify each of his moves.
Write **slide, flip,** or **turn** on each line.

1. _____   2. _____   3. _____

4. _____   5. _____   6. _____

7. _____   8. _____   9. _____

Bonus Box: Draw yourself doing a favorite dance move. Then draw and label the slide, flip, or turn of that move.

Taxi Takeoff

Help students' coordinate-graphing abilities take off!

Purpose: To use coordinates to plot points on a grid

Students will do the following:
- use coordinates to plot points on a grid
- match coordinates with locations on a grid

Materials for each student:
- copy of page 104
- pencil

Vocabulary to review:
- coordinates (ordered pairs)
- plot

Extension activities to use after the reproducible:

- Link coordinate graphing to your current class literature selection with this idea. After students have read or listened to a literature selection, number the x-axis and y-axis of a sheet of one-inch graph paper (pattern on page 168). Then make copies of the programmed graph and distribute them to students. Challenge students to recall locations from the story setting. For each location, instruct each student to draw a point on his graph and use colored pencils to draw a corresponding illustration in the square beside the point. Upon completion of the graph, have each student make a map key for his graph that names the location for each point shown.

- Try this version of tic-tac-toe to reinforce students' coordinate-graphing skills. Provide each pair of students with programmed coordinate cubes (pattern on page 162) and a labeled and laminated grid as shown. To play the game, the players decide who will be X and who will be O. Then, in turn, each player rolls the coordinate cubes and uses a dry-erase marker to make her mark on the corresponding grid point. If the point is already taken, the player's turn is over. The game continues in this manner until one player wins by making three marks in a row. Then, if desired, have players wipe their grids clean to play the game again.

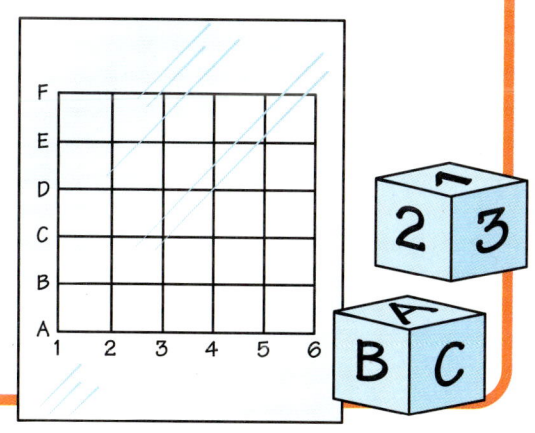

Coordinate graphing

Name _____   Coordinate graphing

Taxi Takeoff

This taxi has many stops on its route.
For route #1 write the coordinates for each location.
For route #2 plot the point for each location and label it on the grid.

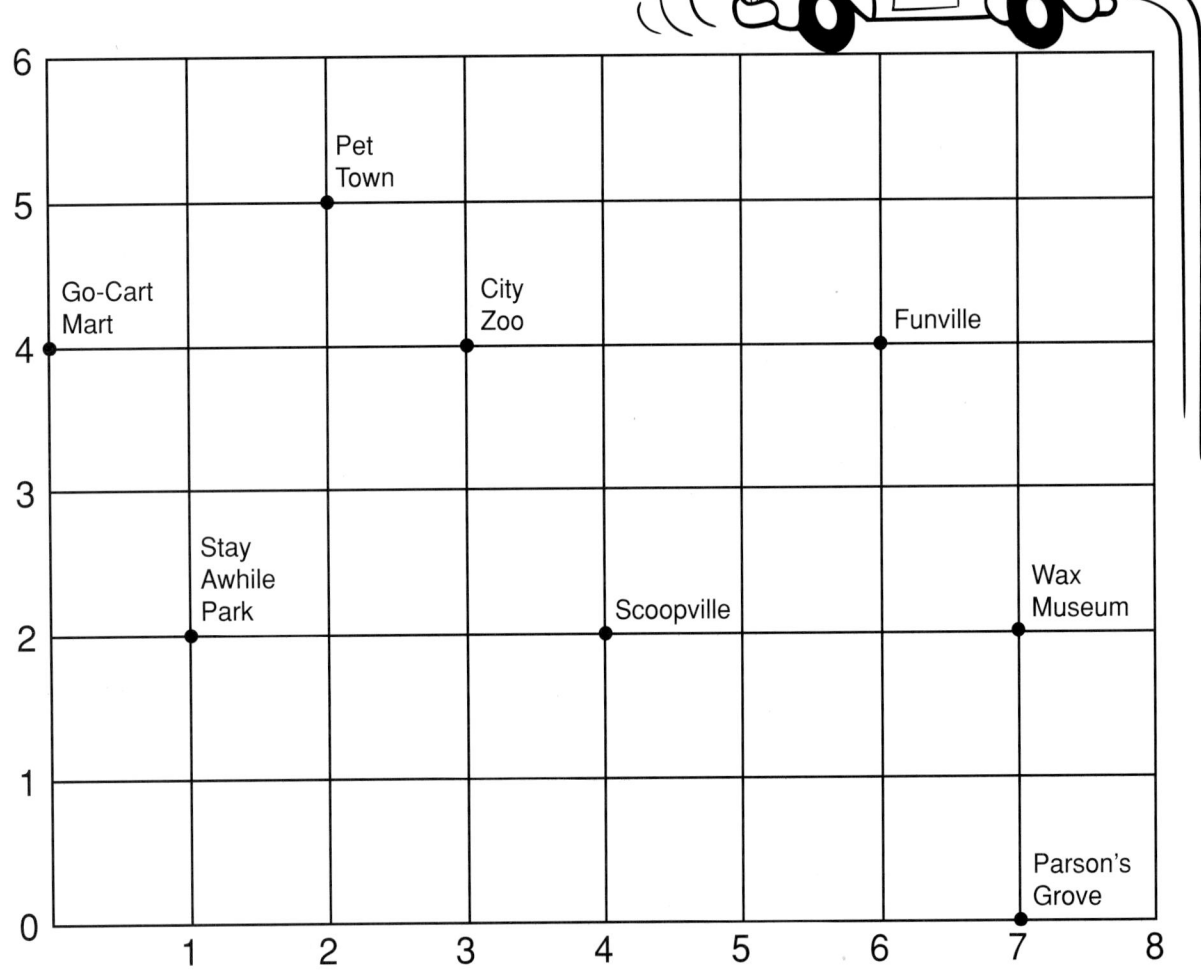

Route #1
1. Funville _____
2. Stay Awhile Park _____
3. Pet Town _____
4. Scoopville _____
5. Parson's Grove _____
6. City Zoo _____
7. Wax Museum _____
8. Go-Cart Mart _____

Route #2
1. Marble Falls (3, 3)
2. Grand Garden (4, 6)
3. Splashdown Pool (5, 1)
4. Jump 'n' Play Station (2, 4)
5. Appleton's Mill (7, 5)
6. Bay Beach (3, 0)
7. Learn-a-Lot School (6, 1)
8. Fun Food Café (5, 5)

Bonus Box: One way to get from Stay Awhile Park to Funville is to travel 5 points east and 2 points north. Name 3 other ways to get from Stay Awhile Park to Funville.

Quilt Quest

Develop students' spatial sense with this quilt-related activity!

Purpose: To develop spatial sense

Students will do the following:
- create shapes using specific types and numbers of pattern blocks
- construct similar figures larger than the size of given shapes
- determine the number of pattern blocks needed by finding the missing addend

Materials for each pair of students:
- copy of page 106
- pencil
- pattern blocks

Vocabulary to review:
- pattern block shapes: orange square, green triangle, yellow hexagon, red trapezoid, blue parallelogram, tan rhombus

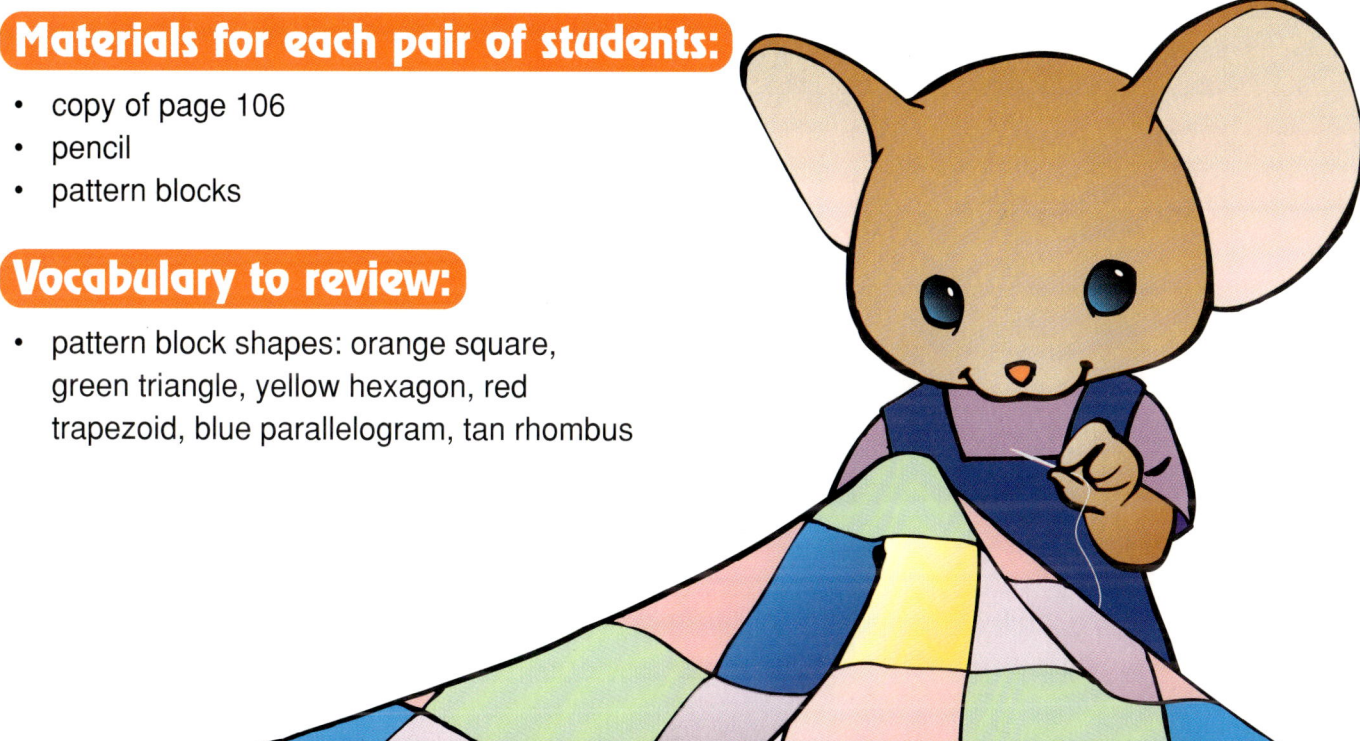

Extension activities to use after the reproducible:
- Reinforce students' spatial sense and recognition of polygons with this pattern-block activity. Provide each student with a set of pattern blocks. Have each student experiment to find different combinations of blocks to form triangles, quadrilaterals, pentagons, hexagons, and octagons. Have students make tracings of their findings to share with the class.

- Develop students' visual and spatial sense with this one-of-a-kind task. Provide each student with a set of pattern blocks and a copy of the dot grid on page 162. Give each student a predetermined amount of time to form a design using yellow hexagons, green triangles, red trapezoids, and blue parallelograms. Then challenge each student to copy his design on the grid and color it in with colored pencils.

Spatial sense

Name _____ Spatial sense

Quilt Quest

Work with your partner to make the shape on each quilt square.
Use the smallest number of pattern blocks possible to form each shape.
Although your pattern-block shape will be larger, it must have the same shape shown.
Under each shape, fill in the blanks to show the number of pattern blocks used.

1.	2.	3.
___ green triangles	___ red trapezoids	___ orange squares
4.	5.	6.
_4_ blue parallelograms ___ tan rhombuses	___ red trapezoids _1_ blue parallelogram	___ green triangles _4_ orange squares
7.	8.	9.
_11_ pattern blocks _8_ green triangles ___ orange squares _1_ tan rhombus	_9_ pattern blocks _1_ orange square _4_ red trapezoids ___ tan rhombuses	_7_ pattern blocks _4_ tan rhombuses ___ red trapezoids _1_ blue parallelogram

Bonus Box: Design your own quilt square. Write the number of pattern blocks used. Tell how many of each shape.

Pep Rally Tally

Cheer students on as they use a tally table to organize and interpret data!

Purpose: To organize and interpret data in a tally table

Students will do the following:
- use tally marks to organize collected data
- complete a tally table
- interpret results from a tally table

Materials for each student:
- copy of page 108
- pencil

Vocabulary to review:
- tally table
- tally marks (tallies)
- data
- results

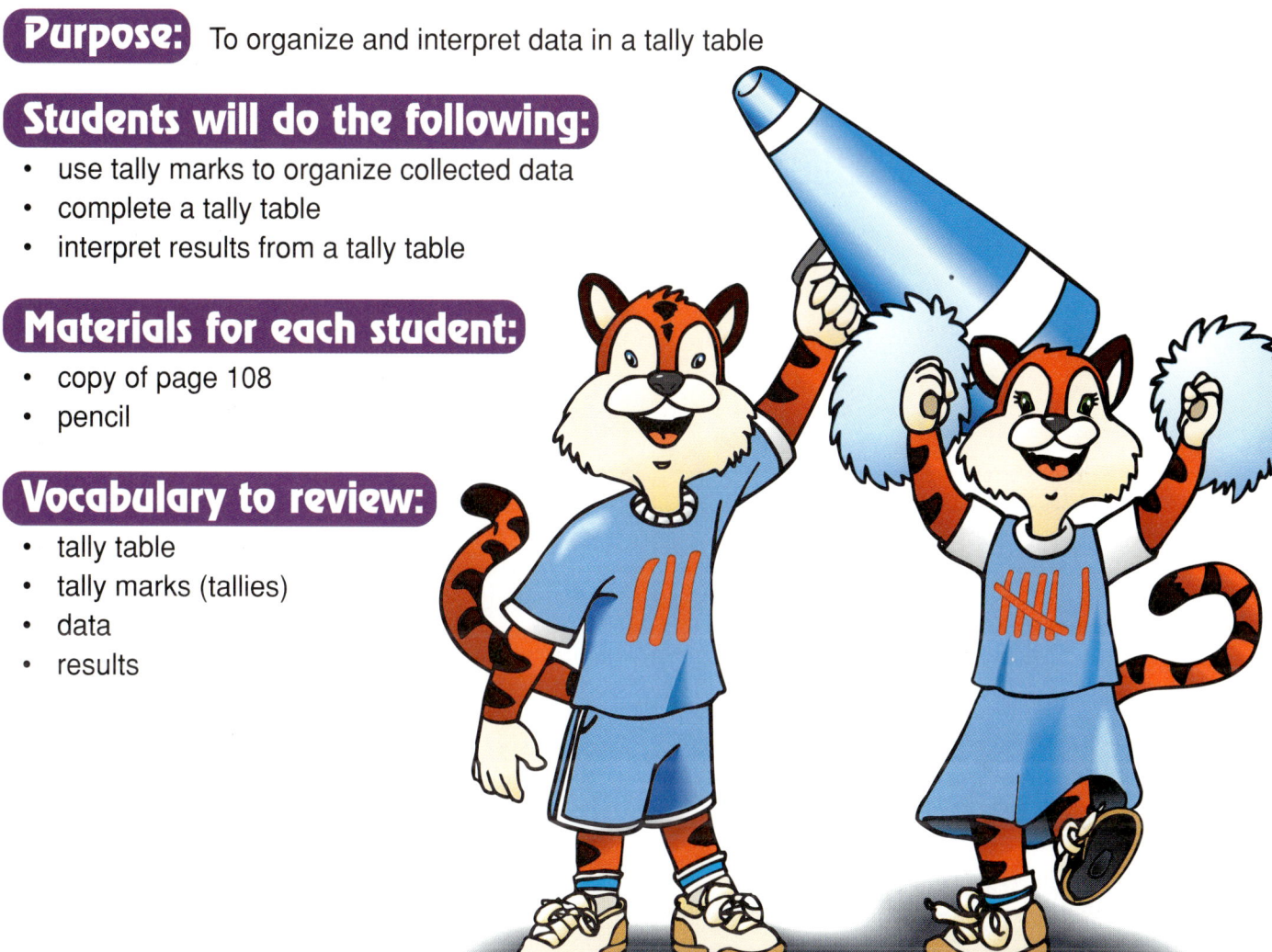

Extension activities to use after the reproducible:
- Show students that organizing and interpreting data in tally tables can be a real treat! Provide each student with a paper towel and a resealable sandwich bag filled with one-half cup of Froot Loops® cereal. Direct each student to sort the contents of his bag by color on the paper towel. Then have him create a tally table to display the data. Remind him to include a title for his table and to label each column. Next, instruct him to write three statements about the results of the data collected. Allow students to enjoy their cereal as they write their interpretations.

- Invite students to learn more about their classmates as they collect data on tally tables. As a class, brainstorm a list of multiple-choice questions that students could ask classmates. Have each student write her chosen question on a sheet of paper, construct a tally table that lists three possible responses, and place her tally table and a pencil on her desk. Organize students to systematically visit each classmate's desk and answer each question by appropriately marking the tally table. After students return to their desks, provide time for them to interpret and share their results.

Organizing and interpreting data

Name _____   *Organizing and interpreting data*

Pep Rally Tally

The Mighty Tigers are having a pep rally before their next game. The cheerleaders have asked the coach to find out which cheer the players like best.
Organize the information from the coach's notes in the tally table. Cross out each player's favorite cheer after you mark it on the tally table.

Coach's Notes About Each Player's Favorite Cheer
V = Victory Growl   W = Winner's Whirl   M = Mighty Tigers

V	W	W	M	V	W	M	V
M	W	V	M	M	V	M	V
V	M	V	W	W	M	M	W
W	M	V	V	M	M	M	V
V	M	V	W	M	V	V	V

Tally Table
Players' Favorite Cheers

Cheer	Tallies
Victory Growl	
Winner's Whirl	
Mighty Tigers	

Tigers, tigers can't be beat!
Get up, stand up, move your feet!
Tigers, tigers on the prowl!
Back off when you hear them growl!
Go, tigers!

Answer each question.

1. How many players like Victory Growl best? _____
2. How many players like Winner's Whirl best? _____
3. How many players like Mighty Tigers best? _____
4. How many players voted for a favorite cheer? _____
5. How many more players like Mighty Tigers than Winner's Whirl? _____
6. List the cheers in order from least to most popular.
   _____
   _____
   _____
7. There are 44 players on the team. How many did not vote for a cheer? _____
8. How many more votes would it take for Mighty Tigers to be the most popular cheer? _____

Bonus Box: The cheerleaders above are chanting the Victory Growl cheer. Create a tally table to show how many times the following words are in the cheer: the word *tigers,* words that rhyme with *eat,* and words that rhyme with *owl.*

Survey Celebration

Show students how to get quick results from surveys!

Purpose: To organize and interpret data collected from a survey

Students will do the following:
- interpret data from a survey
- organize and interpret data on a tally table

Materials for each student:
- copy of page 110
- pencil

Vocabulary to review:
- survey
- responses
- tally marks (tallies)
- data

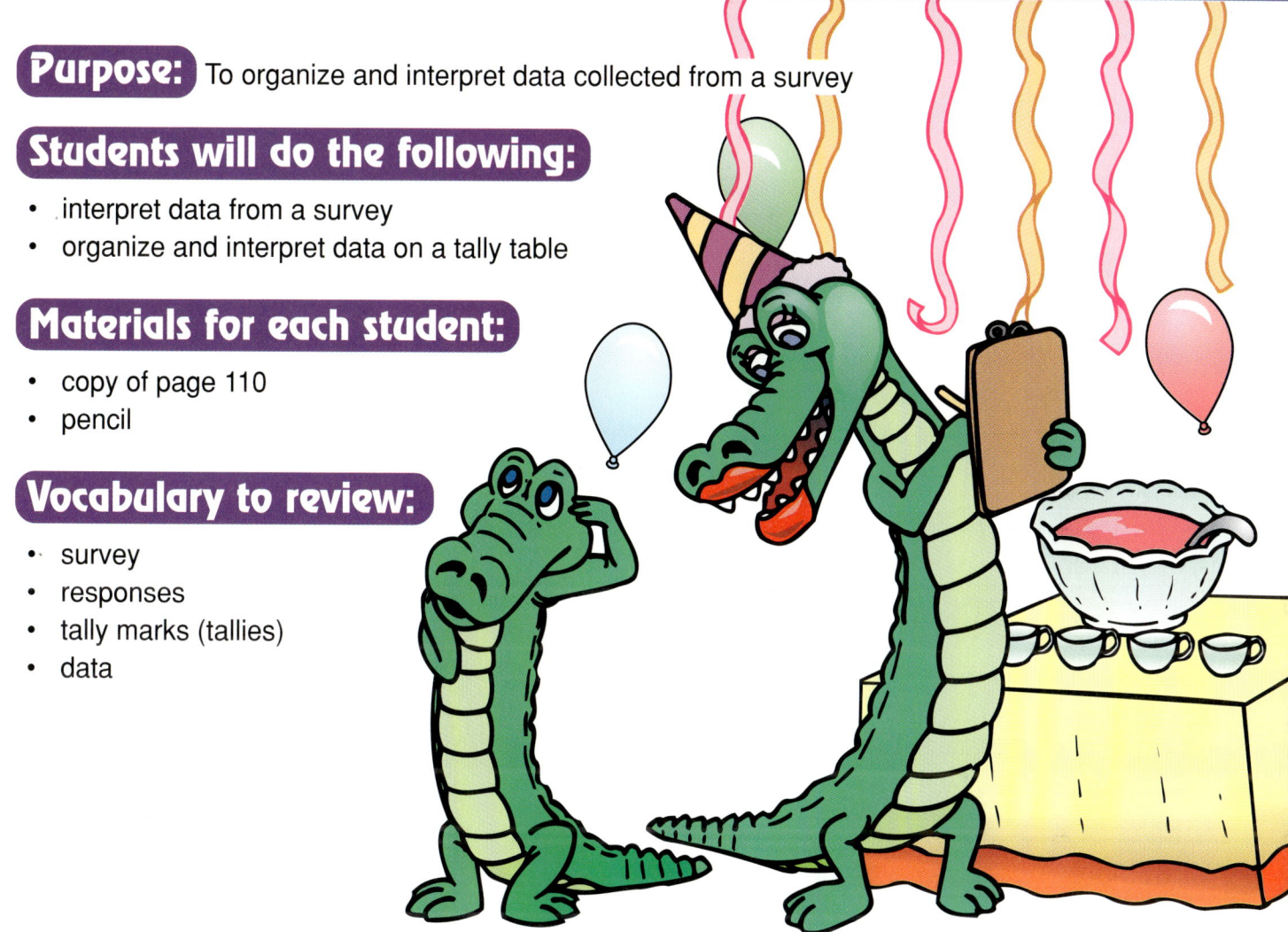

Extension activities to use after the reproducible:
- Help students understand that there are many different ways to conduct a survey. Provide each student with a sticky note. Tell him to write his favorite ice-cream flavor on the note. Invite each student to post his note on the board, grouping like flavors together. Next, have students brainstorm other ways you could have collected the data about their favorite flavors *(distributing a questionnaire or having each student raise his hand to indicate his favorite flavor)*. Then, if desired, have each small group design and conduct its own survey using one of the ways shared.

- Have students use information from a survey to help start planning your next class party. Provide each small group with markers and a sheet of chart paper. Assign a question to each group, such as "Which type of dessert should we serve?" or "When should the party take place?" Instruct each group to design a tally table that includes the question and several answer choices. Post each tally table on the board. Have each student systematically tally his response to each question. Then have each student design a party invitation that reflects the results of the survey.

Organizing and interpreting data

Name _____   *Organizing and interpreting data*

Survey Celebration

Ally Gator took a survey to help plan her party.
She had each of her friends write their favorite ice-cream flavor, drink, and party time on a card.
Study each card and mark each response on the correct tally table.
Then answer the questions below.

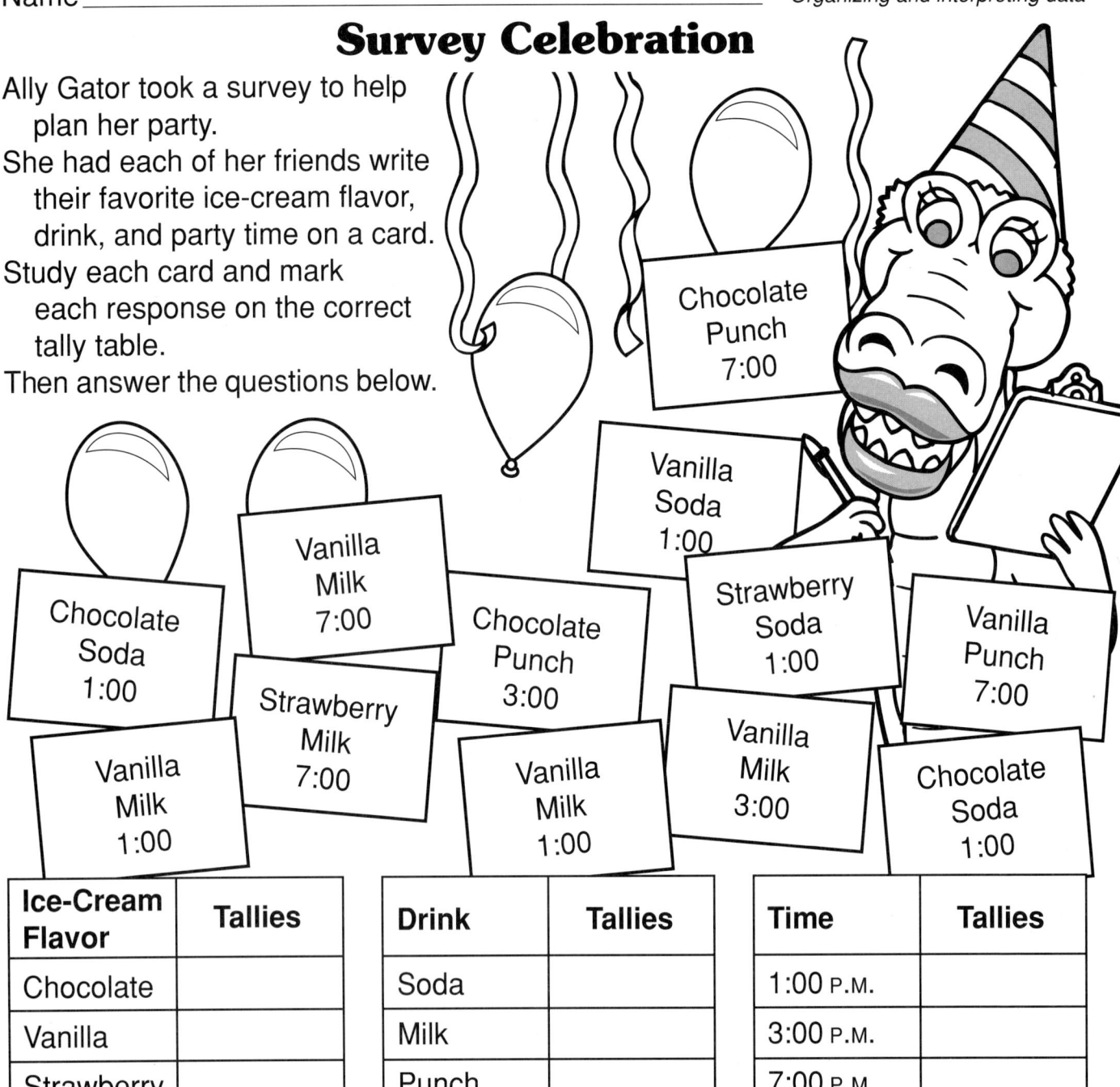

Ice-Cream Flavor	Tallies
Chocolate	
Vanilla	
Strawberry	

Drink	Tallies
Soda	
Milk	
Punch	

Time	Tallies
1:00 P.M.	
3:00 P.M.	
7:00 P.M.	

1. Which ice-cream flavor do Ally's friends like most? _____
2. Which drink do Ally's friends like most? _____
3. Which is the least favorite drink? _____
4. What would be the best time for Ally to have the party? _____
5. Ally only gave her friends 3 choices of ice-cream flavors, drinks, and times. If she would not have listed the choices, how might her results be different?
   _____

Bonus Box: List 3 other activities that a survey could help you plan.

Graph-a-Snack

Help students take a bite out of pictographs with this appetizing activity!

Purpose: To compare and interpret pictographs

Students will do the following:
- interpret data in pictographs
- compare pictographs

Materials for each student:
- copy of page 112
- pencil

Vocabulary to review:
- pictograph
- data

Extension activities to use after the reproducible:

- Here's a newsworthy activity to provide more practice with pictographs! Cut an assortment of headlines from a newspaper. Provide each student with a headline, a glue stick, a sheet of construction paper, and a marker. Instruct him to glue the headline at the top of the construction paper. Then have him count the vowels and consonants in the headline and construct a pictograph to display the data. Remind him to include a title, labels, and a key for the graph.

- Students will love stamping out pictographs at this center! Program a supply of cards with fictional data (see the examples shown). Place the cards, rubber stamps, ink pads, blank index cards, paper, and pencils at a center. At the center, a student selects a programmed card and reads the data. She chooses a stamp to represent the data and assigns the stamp a value of two. She then uses the stamp to construct her pictograph. To show a value of one, she places a blank index card under half of the rubber stamp before stamping it. Finally, she writes a title, labels, and a key for her pictograph. If desired, post the pictographs on a bulletin board titled "Stamping Our Way to Graphing Success!"

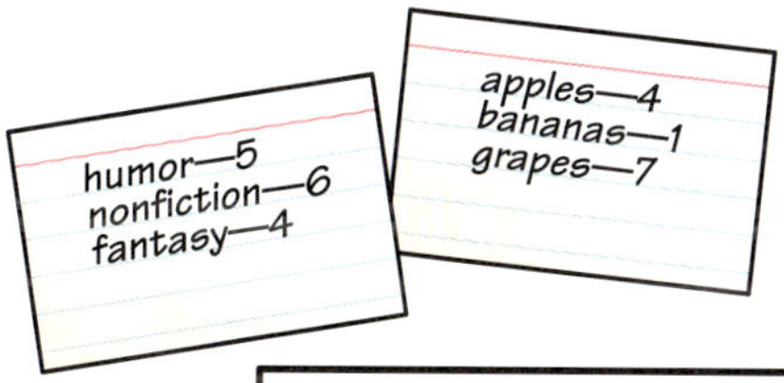

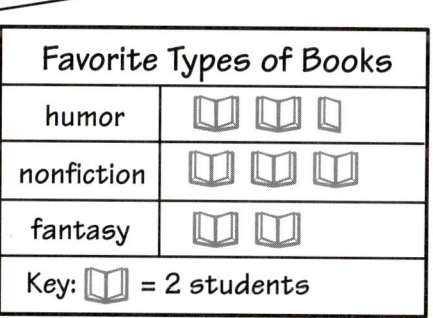

Pictographs 111

Name _____ Pictographs

Graph-a-Snack

Each pictograph shows the amount of nutrients in a single snack bag.
Use the pictographs to answer the questions below.

Protein

Snack	Grams of Protein
Chipper Chips	🐟🐟
Presto Popcorn	🐟
Tasty Tortillas	🐟🐟
Mega Munchies	🐟

Key: 🐟 = 1 gram of protein

Carbohydrates

Snack	Grams of Carbohydrates
Chipper Chips	🍞🍞🍞
Presto Popcorn	🍞
Tasty Tortillas	🍞🍞🍞🍞
Mega Munchies	🍞🍞🍞

Key: 🍞 = 5 grams of carbohydrates

Fat

Snack	Grams of Fat
Chipper Chips	🐷🐷🐷🐷
Presto Popcorn	🐷🐷½
Tasty Tortillas	🐷🐷🐷🐷🐷
Mega Munchies	🐷🐷🐷

Key: 🐷 = 2 grams of fat

1. Which snack has the most fat? _____
   _____ How much?
   _____

2. Which snack has the least carbohydrates?
   _____ How much?
   _____

3. Which 2 snacks have 30 grams of carbohydrates combined? _____

4. How many more grams of fat are in Chipper Chips than in Presto Popcorn?
   _____

5. How many more grams of carbohydrates than protein are in Mega Munchies?
   _____

6. Circle the greater amount:
   grams of fat in Tasty Tortillas
   grams of carbohydrates in Chipper Chips

7. Which snack has 15 grams of carbohydrates and 6 grams of fat?
   _____

8. If you want a snack with the least amount of carbohydrates and the least amount of fat, which snack should you choose?
   _____

Bonus Box: Construct a pictograph to display the data below.
Let ☺ = 10 calories.
Chipper Chips—150 calories
Presto Popcorn—80 calories
Tasty Tortillas—190 calories
Mega Munchies—110 calories

Tuning In to Graphing

Watch students' enthusiasm grow as they tune in to pictographs!

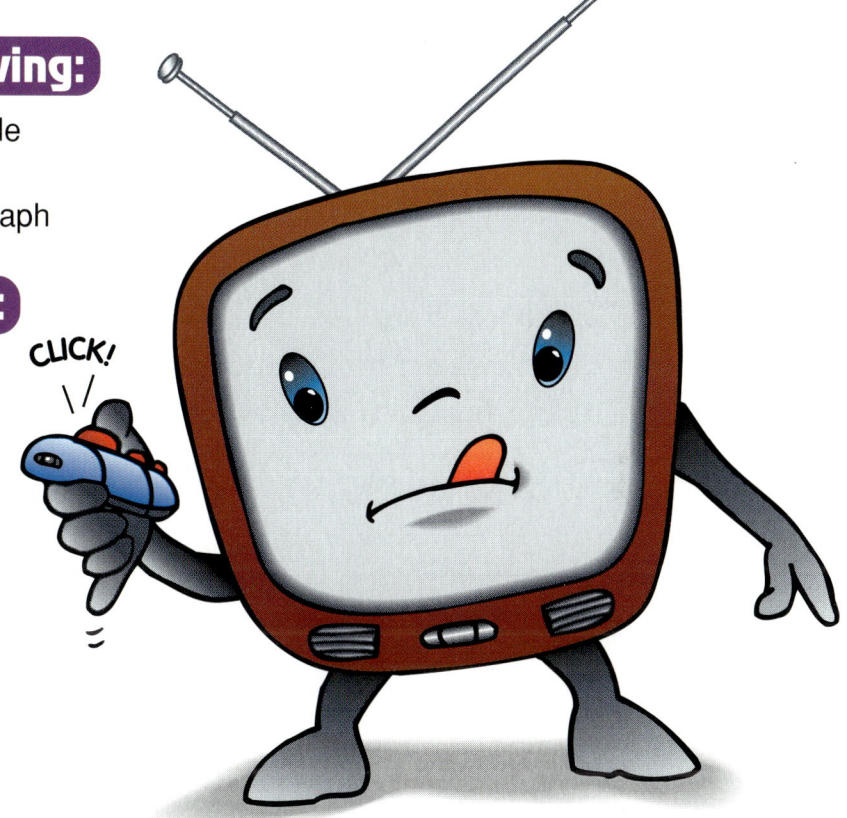

Purpose: To construct and interpret pictographs

Students will do the following:
- interpret data on a television schedule
- construct a pictograph
- interpret data presented on a pictograph

Materials for each student:
- copy of page 114
- pencil

Vocabulary to review:
- pictograph
- data

Extension activities to use after the reproducible:
- Clue youngsters in to pictographs with this problem-solving task! Provide youngsters with the clues listed below, and challenge them to create pictographs that display the information. Direct students to include appropriate titles, labels, and keys for their graphs. Students' pictographs will vary but should show that Jay ate four pieces of candy, David ate five pieces, Kim ate ten pieces, and Lori ate nine pieces.
    Clues:
    - Lori ate one less piece of candy than Kim.
    - Kim ate twice as much candy as David.
    - David ate one more piece of candy than Jay.
    - Jay ate four pieces of candy.
- Provide more practice with pictographs by having your youngsters get the scoop on students in other classes! Have each small group of students brainstorm a question and five choices that could be represented on a pictograph and then create a tally chart for collecting responses to the question. Arrange for each group to pose its question and choices to students in a different class. Have each group record its initial findings on its tally chart. Afterward, challenge the groups to create pictographs to show their results. Direct each group to include a graph title and labels, choose a symbol to use in the pictograph, and determine how many students the symbol should represent. Upon completion of the pictographs, instruct each group to analyze its graph and write five things that can be learned from the graph.

Pictographs

Name _____   Pictographs

Tuning In to Graphing

Study the TV schedule below.
Then complete the pictograph to show how much time each type of show is on.

Channels	MORNING SCHEDULE					
	8:00	8:30	9:00	9:30	10:00	10:30
3	News Today		Cooking With Connie		The Gift of Gab	
5	Animation Station	Mystery of the Lost Homework				
7	Funny Bunny Cartoons	News Update		Inside-out Sports		
11	Cheerleading Competition			Today's Headlines		Perky Pig and Friends

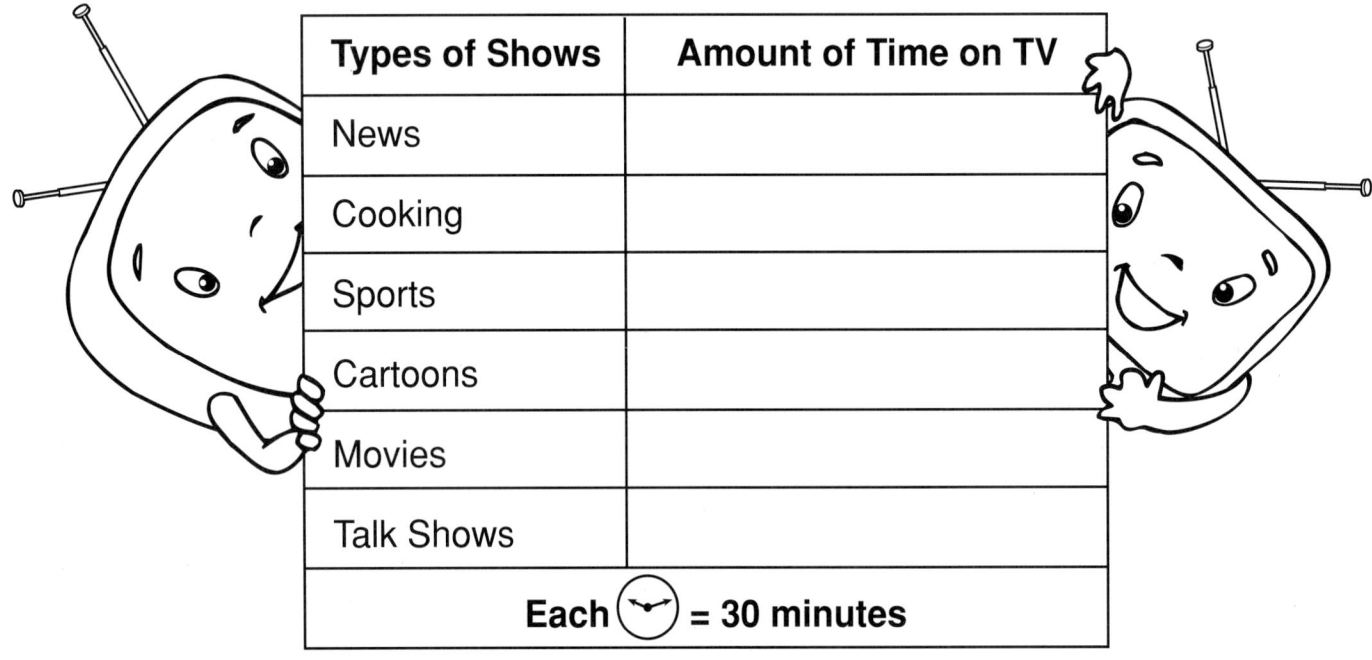

Types of Shows	Amount of Time on TV
News	
Cooking	
Sports	
Cartoons	
Movies	
Talk Shows	

Each 🕐 = 30 minutes

Answer each question below.

1. What would be a good title for the pictograph? _____

2. Which type of show is on the most? _____

3. Which type of show is on the least? _____

4. Which types of shows are on for the same amount of time? _____
   _____

5. How much more time are news shows on than cooking shows? _____

Bonus Box: Write the name of your favorite television show. Tell what day, time, and how the long the program runs.

Gotta Have Heart!

Get to the heart of bar graphs one beat at a time!

Purpose: To construct and interpret bar graphs

Students will do the following:
- construct a bar graph
- compare bar graphs
- interpret data presented in bar graphs

Materials for each student:
- copy of page 116
- pencil

Vocabulary to review:
- bar graph

Extension activities to use after the reproducible:

- To extend the activity on page 116, pair students and have each student in each pair write his name in the first column of the chart beside the star. To complete the second column of the chart, a student in the pair jogs in place for 15 seconds while his partner monitors the time. He then takes his pulse for six seconds and multiplies the number by ten to find his number of heartbeats per minute. After recording the results on his chart, he alternates roles with his partner, who repeats the process in the same manner. The partners continue to alternate roles until each column is complete. After each twosome has collected its data, have each student independently create a bar graph to display his results.

- Copy the tally table shown on the chalkboard. Have students make a bar graph to display the data, giving the graph an appropriate title and labels. Then have students analyze their graphs and write five true statements based on the data.

Favorite Season	Tallies
Winter	|||| |||
Spring	|||| |||| |||| ||
Summer	|||| |||| ||||
Fall	|||| ||||	

Bar graphs

Name _____   Bar graphs

Gotta Have Heart!

Some students wanted to see how jogging can change their heartbeats.
Study the students' results on the chart.
Then complete the bar graphs to show each students' results.

Heartbeats per minute after jogging in place for…				
	15 sec.	30 sec.	45 sec.	60 sec.
Adrian	120	135	140	150
Kelly	115	125	130	145
Brett	100	105	130	140

★

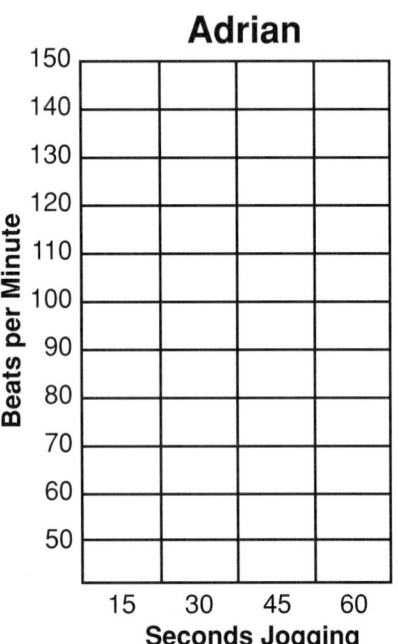
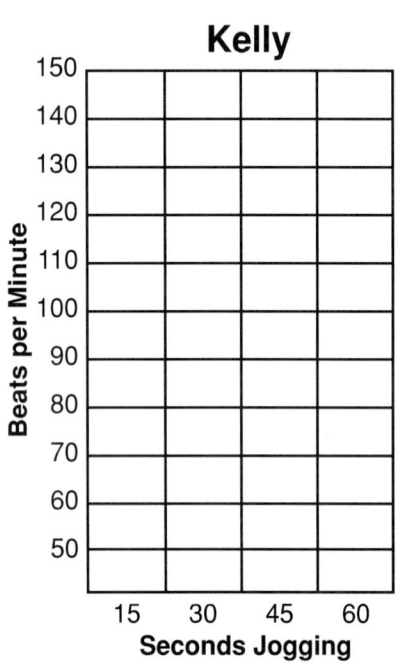
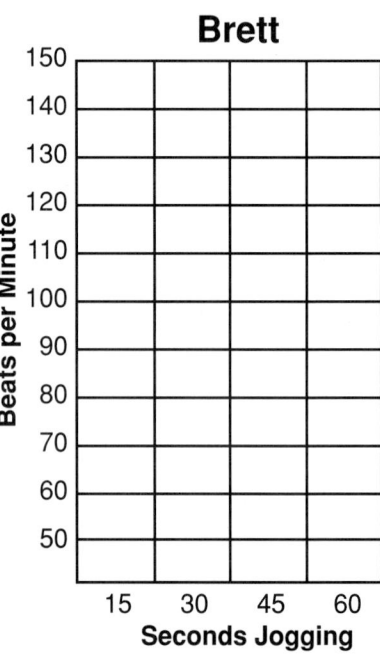

Use the data from the bar graphs to complete each sentence below.

1. Kelly's pulse was _____ than Brett's after jogging 30 seconds.

2. After jogging 15 seconds, _____ had the highest pulse.

3. After jogging 60 seconds, _____ had the lowest pulse.

4. After jogging _____ seconds, Kelly had 15 more beats than Brett.

5. After 15 seconds, Adrian had _____ more beats than Brett.

Bonus Box: List 2 things you can do to make your pulse faster. Then list 2 things you can do to make it slower.

Ready, Set, Recycle!

Help students put graphing skills to good use with this recycling task!

Purpose: To construct and interpret bar graphs

Students will do the following:
- organize data
- construct a bar graph
- interpret a bar graph

Materials for each student:
- copy of page 118
- pencil

Vocabulary to review:
- bar graph
- data

Extension activities to use after the reproducible:
- Link literature and graphing with a survey of students' favorite story characters. On the board, list five characters from stories read in class. Have each student mark a tally beside his favorite character. Next, instruct each student to construct a bar graph based on the data collected. Then have him study his graph and write three related sentences about it.

- Give students a hunger for graphing by having them keep track of favorite cafeteria foods! Tell students that they will conduct a study to find out which cafeteria foods students prefer. Then create a tally table showing a variety of lunch entrees, and have each student tally his preference. After discussing the results, have each student use the data to construct a bar graph and then write a letter to the cafeteria servers to share which foods students love to eat!

Bar graphs

Name _____   Bar graphs

Ready, Set, Recycle!

The students at Robeson Elementary School collected glass, plastic, and newspaper for a recycling project.
The numbers below show how many pounds were collected each day.
Use the information to complete the bar graph.
Then answer each question.

Day 1
Glass—4
Plastic—5
Newspaper—10

Day 2
Glass—4
Plastic—6
Newspaper—3

Day 3
Glass—8
Plastic—6
Newspaper—6

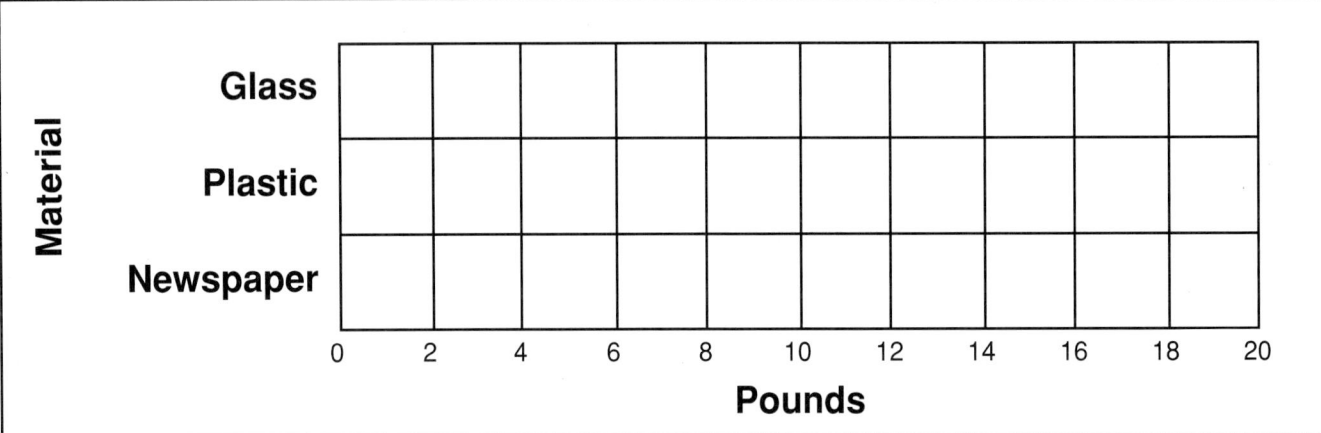

1. Which material was collected the most? _____

2. How many pounds of material were collected all together? _____

3. How many more pounds of newspaper were collected than glass? _____

4. How many pounds of glass and plastic were collected all together? _____

5. Circle the best title for the graph.
    Robeson Elementary School
    Our Favorite Recyclables
    Pounds of Material Collected

6. Circle the question that can not be answered by looking at the graph.
    How many pounds of newspaper were collected?
    How many students collected glass?
    How many more pounds of plastic were collected than glass?

Bonus Box: How many pounds of material were collected all together on Day 1? _____ Day 2? _____ Day 3? _____

118   ©2001 The Education Center, Inc. • Math Skills Workout • TEC3227 • Key p. 174

Splash Into Probability

Dive into probability with this aquatic task!

Purpose: To find probability

Students will do the following:
- make predictions
- conduct a probability experiment
- record data on a tally table
- interpret data from a tally table

Materials for each student:
- copy of page 120
- pencil
- crayons
- scissors
- paper lunch bag

Vocabulary to review:
- probability
- most likely
- least likely
- equal likelihood
- random

Extension activities to use after the reproducible:

- A homework raffle is sure to boost the number of students who turn in assignments! Each day, reward each student who completes all of his homework with a raffle ticket. At the end of a predetermined amount of time, have each student count his tickets and write his name on each one. Lead the class to identify students who have a high chance of winning the raffle. Guide students to discover that those students with the most tickets are more likely to win but are not certain to win. After collecting the tickets, randomly draw a desired number of tickets. Announce each winner and reward him with a prize, such as extra recess or a no homework pass.

- Help students master probability with this thought-provoking partner game. Have each twosome place five red cubes and five blue cubes into a paper bag. Direct each student to make a recording sheet like the one shown. To play, Partner 1 predicts the color of the first cube to be picked, writes her prediction, and then draws a cube. She records the actual color, then gives herself a point if her prediction was correct. Leaving the cube out of the bag, Partner 2 takes a turn. The game continues in this manner until all of the cubes have been drawn. The player with more points wins the round. After a few rounds of play, lead students to discover that the likelihood of choosing a particular color decreases each time the color is drawn.

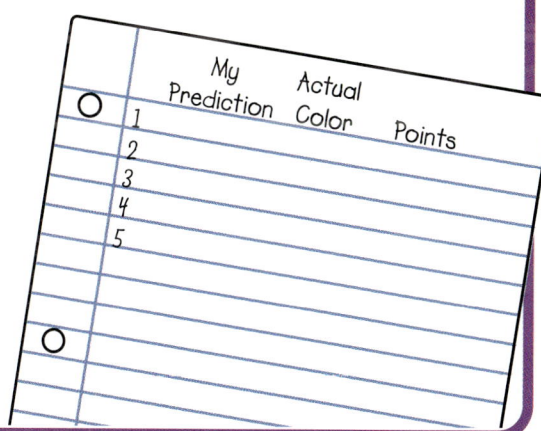

Name _____   Probability

Splash Into Probability

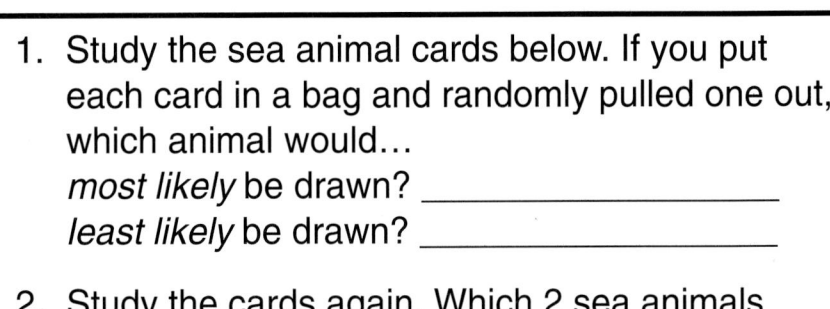

1. Study the sea animal cards below. If you put each card in a bag and randomly pulled one out, which animal would…
   most likely be drawn? _____
   least likely be drawn? _____

2. Study the cards again. Which 2 sea animals would have an *equal likelihood* of being drawn?
   _____

3. Follow the steps to complete a probability experiment.
   Step 1: Color and cut out the sea animal cards below.
   Step 2: Put the cards in your paper bag.
   Step 3: Randomly draw a card. Make a tally mark on the table to show which card you pulled. Then put the card back in the bag.
   Step 4: Repeat this for a total of 25 times.

Sea Animals	Tallies
shark	
starfish	
sea horse	
eel	
jellyfish	

4. Think about your experiment. Which sea animal did you pick most? _____
   _____ Least? _____

5. Would the results of your experiment have been different if there were an equal number of each animal card? _____ Explain your answer. _____
   _____

Bonus Box: If the probability of drawing a starfish were *certain*, which cards would be in your bag? If the probability of drawing a shark were *impossible*, which cards would be in your bag?

©2001 The Education Center, Inc. • *Math Skills Workout* • TEC3227 • Key p. 174

shark	shark	shark	shark	shark	starfish	starfish
starfish	sea horse	eel	eel	jellyfish	jellyfish	jellyfish

A Probability Picnic

Let students spin their way to probability success!

Purpose: To find probability using a spinner

Students will do the following:
- make predictions
- conduct a probability experiment
- record data on a bar graph
- interpret data on a bar graph

Materials for each student:
- copy of page 122
- pencil
- paper clip

Vocabulary to review:
- probability
- most likely
- least likely
- experiment
- possible outcomes

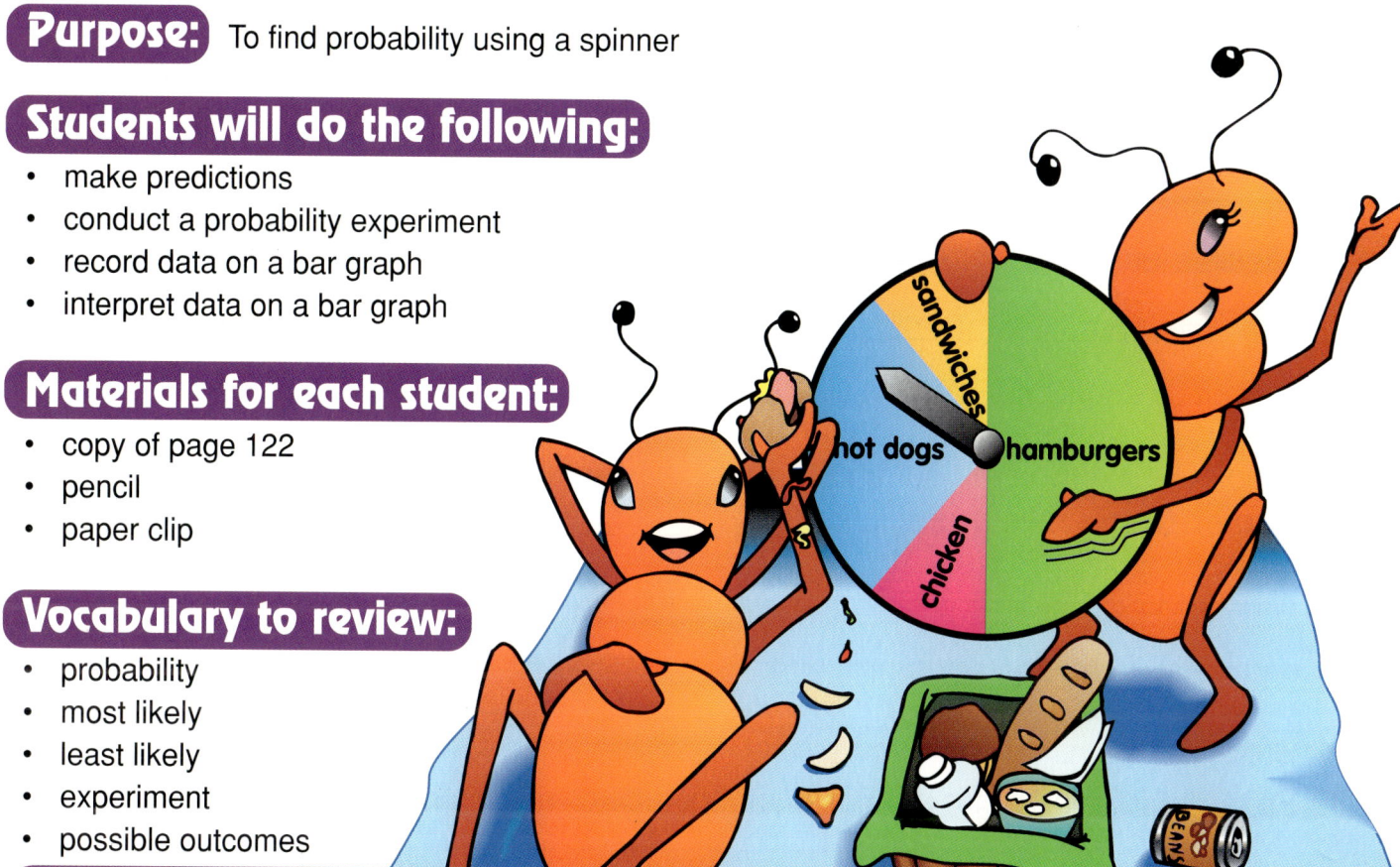

Extension activities to use after the reproducible:

- Extend the activity on page 122 to give youngsters' comparative skills a real spin! Provide each student with a sheet of drawing paper. Have her fold the paper in half twice to create four sections. Next, instruct each student to select her favorite food item from the spinner shown on her copy of page 122, illustrate the item in one section of her paper, and label each of the remaining sections with one of the probable outcomes: most likely, least likely, equally likely. Next, direct each student to create a spinner in each section so that the probability of landing on her food item matches the section label.

- Send students spinning over probability with this activity! Make five enlarged spinners colored as shown (pattern on page 166). If desired, laminate each spinner for durability. Mount the spinners on the chalkboard and number each one. Direct each student to prepare a table similar to the one shown. Challenge him to study each spinner and determine the probability of landing on blue or yellow. Have him write the correct spinner number in each section of the table. Then invite students to share their completed tables.

Probability	Blue	Yellow
Certain		
Likely		
Equally likely		
Unlikely		
Impossible		

Probability 121

Name _____   Probability

A Probability Picnic

Amy Ant and Andy Ant packed so much for their picnic, they can't decide what to eat!
They created a spinner to help them make a decision. Use the spinner to complete the questions and activities below.
To spin the spinner, use your pencil and paper clip.

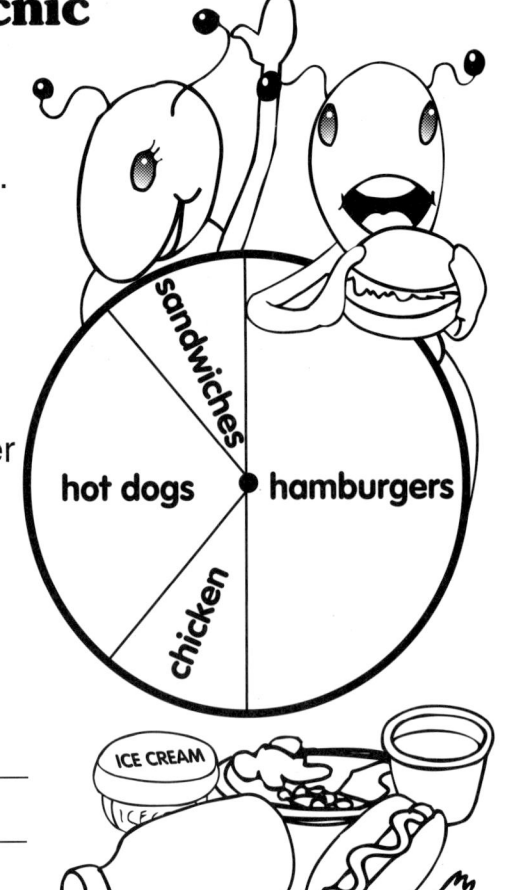

1. Study the spinner. One *possible outcome* is that it could land on hamburgers. What are the 3 other possible outcomes? _____
   _____
   _____

2. Predict which food the spinner is *most likely* to land on. Explain your prediction. _____
   _____
   _____
   _____

3. Spin the spinner 20 times. For each spin, color a box on the graph.

   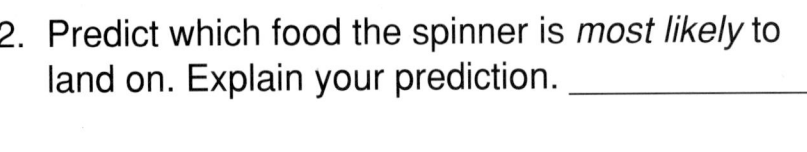
   Hamburgers
   Sandwiches
   Hot dogs
   Chicken

4. Count and record the data you collected in the box below.

   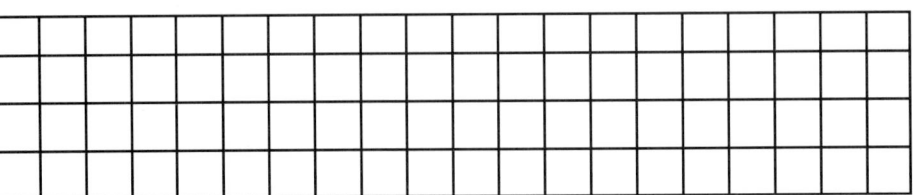
   Hamburgers _____   Sandwiches _____   Hot dogs _____   Chicken _____

5. Compare the results to your prediction. Are they similar? Explain why or why not. _____
   _____
   _____

Bonus Box: Compare your data from above with a classmate's data. Describe what you discover.

Professor Probability

Help students become wise about predicting probability!

Purpose: To predict probability

Students will do the following:
- identify possible outcomes
- make predictions
- record data in a frequency table
- interpret data from a frequency table

Materials for each student:
- copy of page 124
- pencil

Vocabulary to review:
- possible outcomes
- likelihood
- most likely
- least likely
- equally likely
- frequency table
- experiment

Extension activities to use after the reproducible:

- Probability is in the bag with this activity! Construct a frequency table similar to the one shown. Have each student write his favorite subject on a card. Collect the cards and place them in a paper bag. Circulate the bag and have each student draw a card, announce the subject, and then return the card to the bag. Record the results on the table. Challenge students to predict the order of the subjects from most favorite to least favorite based on the data. *(Students should conclude that the subject with the highest frequency is probably the subject that most students prefer.)* To check the predictions, sort and count the cards by subject.

Subject	Tallies	Frequency (total number of tallies)
Language Arts		
Math		
Science		
Social Studies		

- Combine a sweet treat with probability practice. Provide each student with a paper cup, a paper towel, and 20 M&M's® candies. Have her sort and count the candies by color and then construct a frequency table to record the results. Next, have her predict the color that is most likely and least likely to be picked in 20 attempts. Have her place the candies in the cup and test her predictions. Remind her to return the candy to the cup after each draw. Provide time for students to discuss whether their predictions were correct as they enjoy the treat!

Probability

Name _____ Probability

Professor Probability

Professor Probability is conducting an experiment! First, he will place the number cards shown into a hat. Then he will randomly draw a number 20 times, placing the number back in the hat after each draw.

1. Each time Professor Probability chooses a card, how many *possible outcomes* are there? _____

2. How many numbers are less than 20? _____

3. How many numbers are greater than 20? _____

4. Using the words on Professor Probability's hat, predict the likelihood of…

   a. drawing a number less than 20 _____

   b. drawing a number greater than 20 _____

   c. drawing a number less than 41 _____

   d. drawing the number 25 _____

   e. drawing an even number _____

5. Look at the results of the professor's experiment. Circle and count the numbers less than 20 and write the total in the frequency table. Next, underline and count the numbers greater than 20 and write the total in the frequency table.

Results				
14	40	33	33	23
27	33	40	18	40
18	23	18	40	27
33	18	27	27	18

Frequency Table	
Number	Frequency
less than 20	
greater than 20	

6. Now compare the results shown on the frequency table to your answers to questions 4a and 4b above. Were your predictions correct? _____ Explain.

_____

Bonus Box: Repeat Professor Probability's experiment! Write the numbers from the cards shown on 6 slips of paper and randomly draw a number 20 times. Remember to place each number back in the pile before you pick the next number. Record your results on a frequency table.

What Are the Odds?

Get students on a roll with understanding fairness with this probability task!

Purpose: To determine fairness of a game

Students will do the following:
- identify possible outcomes
- determine the fairness of a game
- complete a tally table
- record data on a tally table
- write to explain thinking

Materials for each pair of students:
- copy of page 126
- pencil
- die

Vocabulary to review:
- fair
- possible outcomes
- equally likely
- odd/even

Sum	Tallies
1	
3	
5	
8	
9	
12	

Extension activities to use after the reproducible:

- Chances are that students will love playing this probability partner game! Give each twosome a pair of dice and two sheets of paper. Instruct each partner to draw a tally table as shown. In turn, each player chooses a sum from 1 to 12 until all have been chosen. To begin, Player 1 rolls the dice and adds the numbers shown. If the sum is one that she chose, she makes a tally mark on the table. If not, then her partner makes a tally mark. Alternate play continues until one player rolls a sum ten times. After a few rounds of play, have students discuss which sums are more likely to be rolled and why. *(Sums six, seven, and eight are more likely to be rolled because there are more ways to roll them.)* If desired, have students play the game again, using what they learned to select numbers more carefully.

- Have students further explore probability with personalized dice! In advance, make a copy of the die pattern on page 162 for each group of six students. Before assembling each die, label each section with a student's name in the group. (If a group has less than six students, leave that section of the die blank.) Challenge each group to study its die and predict the probability of rolling a girl's (or boy's) name as being certain, likely, equally likely, unlikely, or impossible. Direct each student to record his group's predictions and make a tally table for recording results. Instruct group members to take turns rolling the die until it has been rolled 20 times. Then have each student compare his final results to his predictions. For further exploration, have students determine the probability for rolling the name of a student who is a bus rider, went to a different school the previous year, or has a younger sibling.

Probability

Names _____   *Probability*

What Are the Odds?

The Fair Game toy company is working on a new game called What Are the Odds?
Play the game according to the directions.
Then answer the questions below.

Remember, a game is fair when each player has an equal chance of winning!

Tally Table—Number Rolled	
Players	Tallies
Player 1 (Odd)	
Player 2 (Even)	

Directions for 2 players:
1. Decide who will be Player 1 and Player 2.
2. In turn, roll the die.
3. If an odd number is rolled, Player 1 makes a tally mark on the table. If an even number is rolled, Player 2 makes a tally mark.
4. After a total of 20 rolls, the player with the most tally marks wins the game.

1. One *possible outcome* for rolling the die is rolling the number 1. List the other possible outcomes. _____

2. How many even numbers are on the die? _____ How many odd numbers? _____

3. Do you think that the game is fair? _____ Explain why or why not.
   _____

4. What if the toy company decides to create a new die for the game using the following numbers: 2, 3, 4, 6, 7, 8, and 9. Is this a fair die? _____
   Explain why or why not. _____

5. Create 2 new dice for the game: 1 fair and 1 unfair. List the 6 numbers for each die. _____

Bonus Box: Create a die that would be unfair for Player 1. Then create a die that would be unfair for Player 2.

Penguin Paintings

Let Pablo Penguin put a little pizzazz into understanding patterns!

Purpose: To identify patterns

Students will do the following:
- identify patterns
- write to describe patterns

Materials for each student:
- copy of page 128
- pencil
- crayons

Vocabulary to review:
- pattern

Extension activities to use after the reproducible:
- Students will line up to participate in this activity! Select several students to come to the front of the classroom and arrange them in random order or in a boy-girl pattern (such as *boy, boy, girl* or *girl, boy, girl*). Then challenge the remaining students to study the lineup and decide whether a pattern is present. If students determine that there is a pattern, have them state the pattern rule. Repeat this activity as often as desired using a different student lineup each time.

- Engage your youngsters in hands-on pattern practice. Provide each student with manipulatives, such as Unifix® cubes, pattern blocks, or uncooked pasta shapes. Using the same materials distributed to students, create a variety of patterns one at a time in front of the room. Each time, have students use the manipulatives at their desks to extend the pattern.

Patterns 127

Name _____   Patterns

Penguin Paintings

Check out Pablo Penguin's cool paintings!
If the painting shows a pattern, write a sentence to describe the pattern.
If the painting does not show a pattern, circle it.

1. Parade of Friends

2. Family Portrait

3. Icebergs at Dawn

_____

_____

4. Dinner Delight

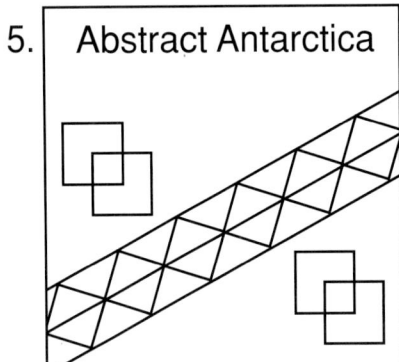

5. Abstract Antarctica

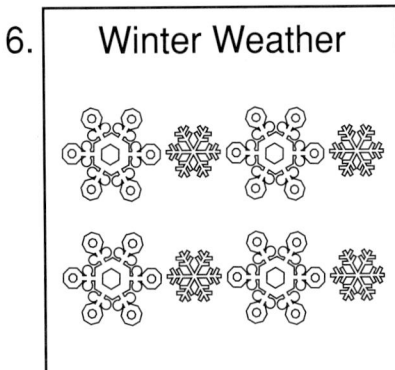

6. Winter Weather

_____

_____

7. Danger at Sea

8. Penguin Dreams

Bonus Box: Draw a picture of something you might see where you live. Include a pattern in your picture.

_____

_____

Flopsy's Florals

Have youngsters visit Flopsy's Florals to practice pattern arrangements!

Purpose: To identify and continue patterns

Students will do the following:
- identify and continue numeric patterns
- identify and continue picture and shape patterns

Materials for each student:
- copy of page 130
- pencil

Vocabulary to review:
- pattern

Extension activities to use after the reproducible:
- Draw figures and record the number of dots in each figure on a transparency as shown. Display the figures and direct students to extend the pattern by illustrating the fifth and sixth figures in the pattern. Have students predict the number of dots needed to draw the seventh and eighth figures. *(28 and 36)* Direct students to draw the figures to verify their predictions. Then instruct each student to describe the rule of the pattern in writing. Repeat this activity using dots to form different figures in a pattern sequence.

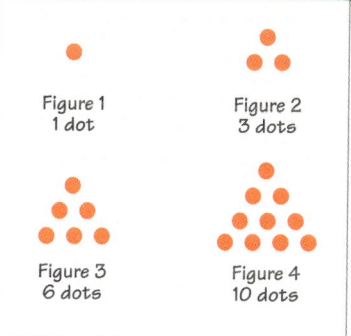

Figure 1 — 1 dot
Figure 2 — 3 dots
Figure 3 — 6 dots
Figure 4 — 10 dots

- Have students display their knowledge of geometric and number patterns with tables. On the chalkboard, draw a table like the one shown. Then, with students' assistance, complete the first five columns of the table. Challenge students to predict the information for the remaining columns. If desired, have students draw models to verify the information on the table. Finally, have students identify number patterns created on the table and solve additional questions, such as "How many sides would be on 20 squares?" Repeat this process as often as desired using a different polygon shape each time.

Number of Squares	1	2	3	4	5	6	7	8	9	10
Number of Sides										

Flopsy's Florals

Flopsy's Florals is having a special sale! Continue the number patterns to complete the price charts.

Flower Prices

Roses

Number	2	4	6			16	
Price	$4	$8	$12		$24		

Carnations

Number	5	10	15			35	
Price	$3	$6	$9				$27

Now draw a picture or shape to continue each pattern.

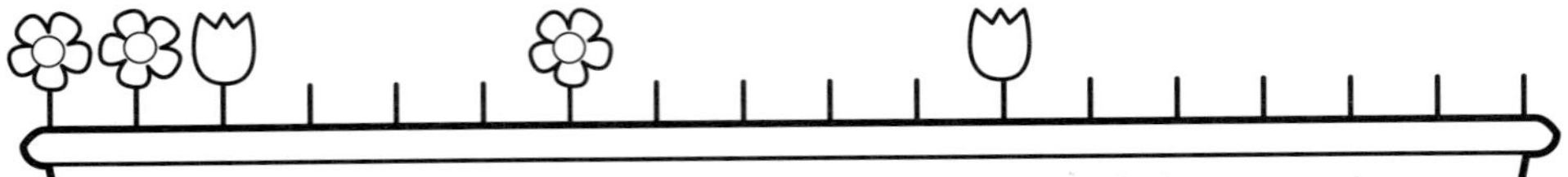

Bonus Box: Pretend the first flower in this box is red, the second flower is yellow, and the third flower is orange. If this pattern continues, what color will the last flower be?

The Case of the Missing Addends

Put your super sleuths on the trail of detecting missing addends!

Purpose: To find missing addends

Students will do the following:
- determine the missing addends
- add to check answers

Materials for each student:
- copy of page 132
- pencil
- crayons

Vocabulary to review:
- addend
- sum

Extension activities to use after the reproducible:

- Help students take the mystery out of missing addends! Provide each student with crayons, scissors, and a copy of the centimeter graph paper on page 167. Write a number sentence with a missing addend on the board. Instruct each student to cut a row of graph-paper squares equal to the sum. Then have him color the number of squares that represent the given addend. To find the missing addend, he counts the uncolored squares. Then he glues each row of squares to a sheet of construction paper and writes a matching number sentence. Have students solve additional problems in the same manner.

- Put students on the path to finding missing addends with this independent free-time activity. Fold ten sheets of construction paper in half. Keeping the paper folded, program the front flap of each sheet as shown. Then lift each flap and program it with the information shown under each sample. Mount the folded signs randomly around your classroom. When a student has free time, she goes to the START sign, lifts the flap, and records the problem. After determining the missing addend, she goes to the sign that displays the correct answer. She lifts the flap of that sign to find a new problem to solve. The student continues this process until she reaches the last sign. If desired, reward each student who completes this free-time task with a small treat.

START	1	2	3	4
☐ + 7 = 11	7 + ☐ = 13	☐ + 3 = 11	☐ + 0 = 5	8 + ☐ = 15

5	6	7	8	9
☐ + 6 = 7	☐ + 8 = 10	9 + ☐ = 12	☐ + 7 = 16	You're done. Great job!

Missing addends 131

Name_____   Missing addends

The Case of the Missing Addends

Detective Dawg is hot on the trail!
Read each problem.
Write the missing addend in the 🔍.
Add to check your answer.
Then color the matching pawprint.

A. ◯
   + 8

    17

B.   5
   +◯

    11

C.   6
   +◯

    14

D. ◯
   + 3

    10

E.   3
   +◯

    8

F. ◯
   + 6

    12

G.   7
   +◯

    8

H. ◯
   + 9

    13

I.   2
   +◯

    10

J. ◯
   + 7

    14

K.   3
   +◯

    12

L. ◯
   + 6

    13

M. ◯
   + 9

    19

N.   5
   +◯

    14

O. ◯
   + 8

    13

P.   7
   +◯

    9

Q. ◯
   + 5

    9

R. ◯
   + 7

    15

S.   6
   +◯

    13

T.   4
   +◯

    11

Bonus Box: Can understanding subtraction help you solve the problems on this page? Tell why or why not.

What's the Scoop?

Treat students to finding missing factors with this cool activity!

Purpose: To find missing factors

Students will do the following:
- determine missing factors in multiplication facts
- multiply to check answers

Materials for each student:
- copy of page 134
- pencil
- scissors
- glue

Vocabulary to review:
- factor
- product

Extension activities to use after the reproducible:
- Try this partner game for extra practice with missing factors. Provide each student with six index cards. Instruct her to program one card with a multiplication equation that has a missing factor of one. Then have her program the next card so that the missing factor is two. Have her repeat this pattern with the remaining cards. Next, pair students and provide each pair with a number cube. Tell the twosome to stack its combined cards faceup. To play the game, each player in turn looks at the equation shown on the top card, determines the missing factor, and then rolls the cube. If she rolls the correct number, she takes the card. If not, she places the card at the bottom of the deck. Play continues until all the cards have been taken. The player with more cards wins!

- Give students a double scoop of missing factors with this cool center! Cut 16 ice-cream scoops and eight ice-cream cones from construction paper. Program two sets of scoops and cones as shown. If desired, laminate the cutouts for durability. Store each set in a resealable plastic bag. Place the bags in a center. The student removes the cutouts from a bag and assembles four ice-cream cones, each showing a correct multiplication sentence. She returns the cutouts to the bag and then repeats the activity with the other set.

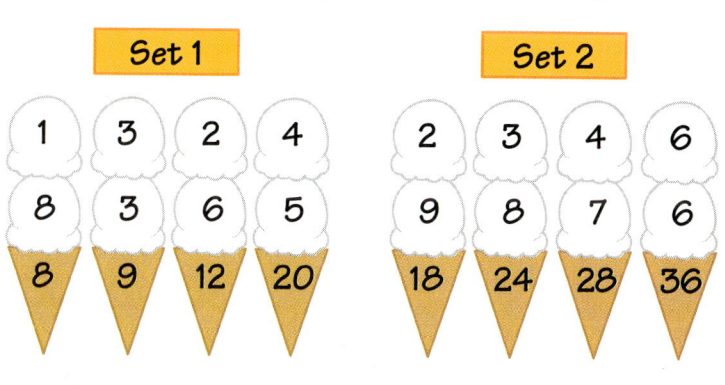

Missing factors

Name _____   Missing factors

What's the Scoop?

Cut out the ice-cream scoops below.
Look at each multiplication problem. Find the missing factor.
Glue the matching scoop in place.
Multiply to check your answer.

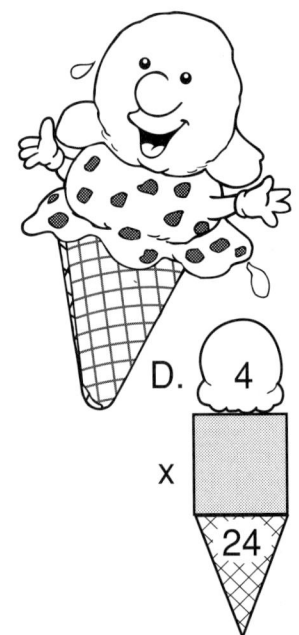

A. 8 × ☐ = 24

B. 9 × ☐ = 72

C. 6 × ☐ = 12

D. 4 × ☐ = 24

E. 9 × ☐ = 45

F. 3 × ☐ = 27

G. 8 × ☐ = 56

H. 7 × ☐ = 7

I. 5 × ☐ = 20

J. 4 × ☐ = 0

K. 6 × ☐ = 24

L. 7 × ☐ = 42

M. 5 × ☐ = 35

N. 9 × ☐ = 54

Bonus Box: Circle the ice-cream treats whose factors have a sum of 10 and a product of 24.

©2001 The Education Center, Inc. • Math Skills Workout • TEC3227 • Key p. 175

| 9 | 6 | 2 | 5 | 6 | 7 | 3 |
| 1 | 7 | 8 | 6 | 4 | 0 | 4 |

134

Cool Calculations

Increase students' understanding of the commutative property with this "penguin-perfect" task!

Purpose:
To understand commutative properties of addition and multiplication

Students will do the following:
- solve addition, subtraction, and multiplication facts
- discover that the commutative property only applies to addition and multiplication
- rewrite problems using the commutative property
- complete sentences that explain the commutative property

Materials for each student:
- copy of page 136
- pencil

Vocabulary to review:
- commutative property
- equation

Extension activities to use after the reproducible:

- Have students explore the commutative property of addition using three addends. Provide each student with a copy of the number cards on page 166. Have each student cut out his cards and stack them facedown. The student pulls three cards and uses the numbers as addends to write a problem to solve. Then he manipulates the order of the numbers as many different ways as possible, solving the problem each time. Have the student shuffle and restack his cards to repeat this process a predetermined number of times. If desired, have students use this same method to explore the commutative property of multiplication with three factors.

- Invite youngsters to advertise their knowledge of the commutative property. Supply each student with a 9" x 12" sheet of construction paper and challenge her to create a slogan that reflects the principle of the property. Then have her write and illustrate the ideas to create a poster (see the example shown). Invite each youngster to share her advertisement before taking it home to share with family members and friends.

Commutative property

Name _____ Commutative property

Cool Calculations

Read about the commutative property.
Then solve each equation.
If the commutative property can be used,
  show it by rewriting the equation.
The first one has been done for you.

Commutative Property
The commutative property states that switching the order of numbers in an equation does not change the answer.
Example:
7 + 2 = 9
2 + 7 = 9

A. 4 x 7 = __28__
   7 x 4 = 28

B. 6 + 3 = _____

C. 5 x 7 = _____

D. 3 x 8 = _____

E. 3 + 2 = _____

F. 4 − 3 = _____

G. 7 + 8 = _____

H. 9 − 6 = _____

I. 6 − 2 = _____

J. 5 x 2 = _____

K. 9 x 4 = _____

L. 5 + 2 = _____

Think about the commutative property.
Then study your answers to the equations above.
Use the information you have learned to answer each question.

1. Does the commutative property work with subtraction? Explain why or why not.
_____

2. Does the commutative property work with addition? Explain why or why not.
_____

3. Does the commutative property work with multiplication? Explain why or why not.
_____

Bonus Box: How can understanding the commutative property help you learn your addition and multiplication facts?

Arithmetic Roundup

Round up your wranglers for practice solving equations with parentheses! Yee-haw!

Purpose: To solve problems that contain parentheses

Students will do the following:
- solve problems that contain parentheses
- insert parentheses to complete equations

Materials for each student:
- copy of page 138
- pencil

Vocabulary to review:
- parentheses
- equation

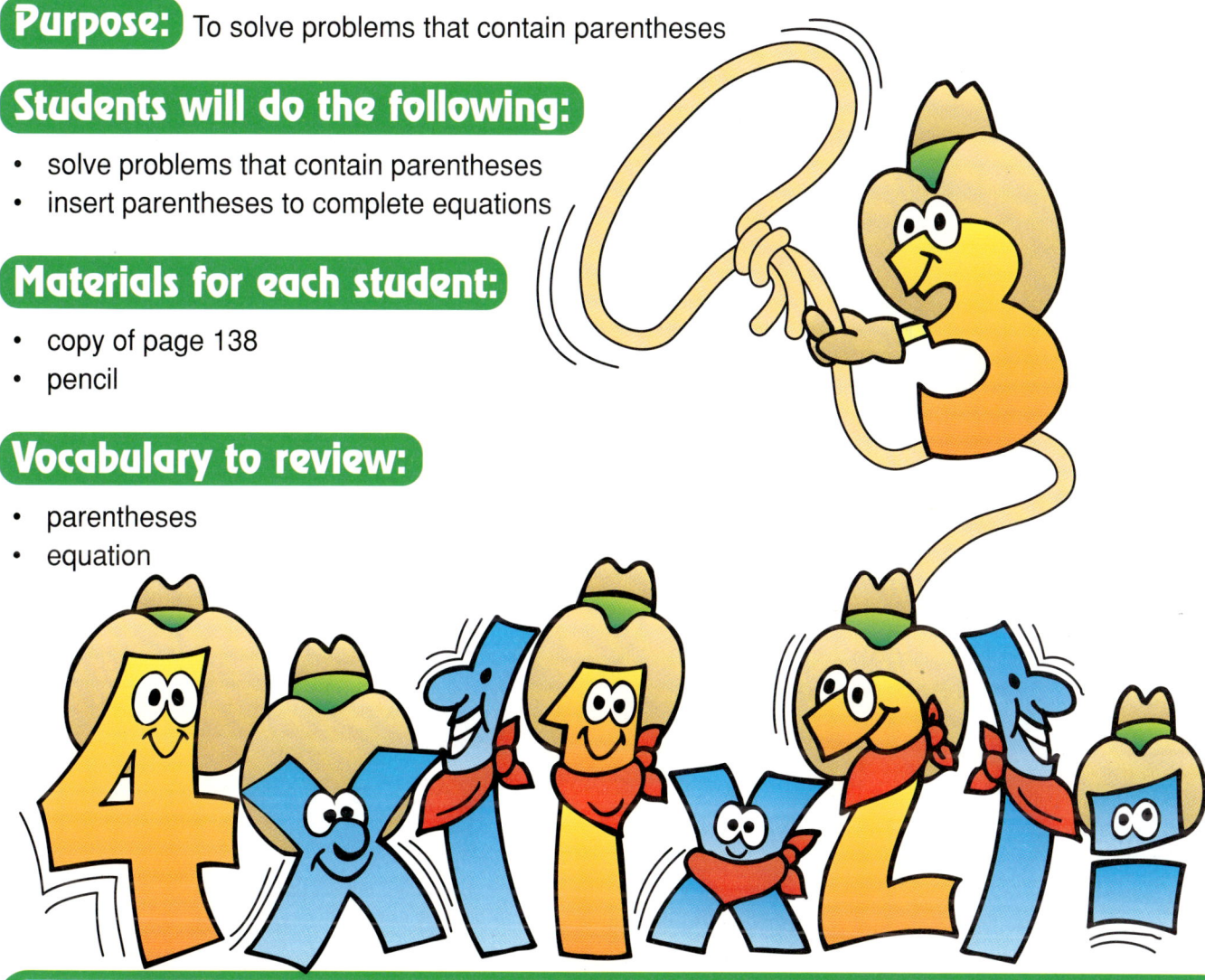

Extension activities to use after the reproducible:
- Reinforce students' understanding of using parentheses by involving them in living equations! Supply each student with an index card and have her write a single-digit number on the card. Then invite three students to stand in front of the chalkboard spaced apart to display their numbers. To form a problem, place parentheses and computation symbols between students by writing them on the chalkboard. Then choose a student volunteer to solve the problem. The first student to correctly respond gets to include her number in the next equation.

- Use this problem-solving activity to provide students with more practice using parentheses in equations! Write a problem on the chalkboard containing parentheses, such as 2 x (3 + 4) = ___. Solve the problem as a class. Next, create and discuss possible word problems represented by the equation, such as "Sue reads three books and four magazines two times a week. How many items does Sue read each week?" *(14)* Next, divide the class into pairs. Have each pair create and write an equation with parentheses and a corresponding word problem. Allow time for student pairs to share their word problems with the class. Encourage their classmates to give them constructive feedback.

Using parentheses

Name _____   Using parentheses

Arithmetic Roundup

Howdy, partner! Solve each problem.
Remember that the part in parentheses must always be solved first!

A. 5 + (8 x 2) = ____   B. 7 x (5 – 1) = ____   C. (1 + 7) x 7 = ____

D. (4 – 2) + 6 = ____   E. (5 x 3) + 3 = ____   F. (2 + 3) x 4 = ____

G. 1 + (7 x 7) = ____   H. (2 x 5) – 4 = ____   I. 6 – (1 + 5) = ____

J. 4 – (6 x 0) = ____   K. 4 x (1 + 2) = ____   L. 8 + (3 x 2) = ____

Read each problem.
Draw parentheses around the correct numbers to make the answer true.
The first one has been done for you.

M. (5 – 3) x 2 = 4   N. 2 x 6 + 5 = 17   O. 4 x 7 – 3 = 16

P. 3 + 9 x 3 = 36   Q. 4 + 4 x 3 = 16   R. 9 + 1 x 3 = 30

S. 7 x 4 – 1 = 21   T. 7 x 1 + 3 = 28   U. 4 + 3 x 2 = 10

Bonus Box: Use parentheses to show 2 different ways to write the problem 4 x 1 + 5 = ____. Solve each problem.

Function Junction

Put your youngsters on the train to function table success!

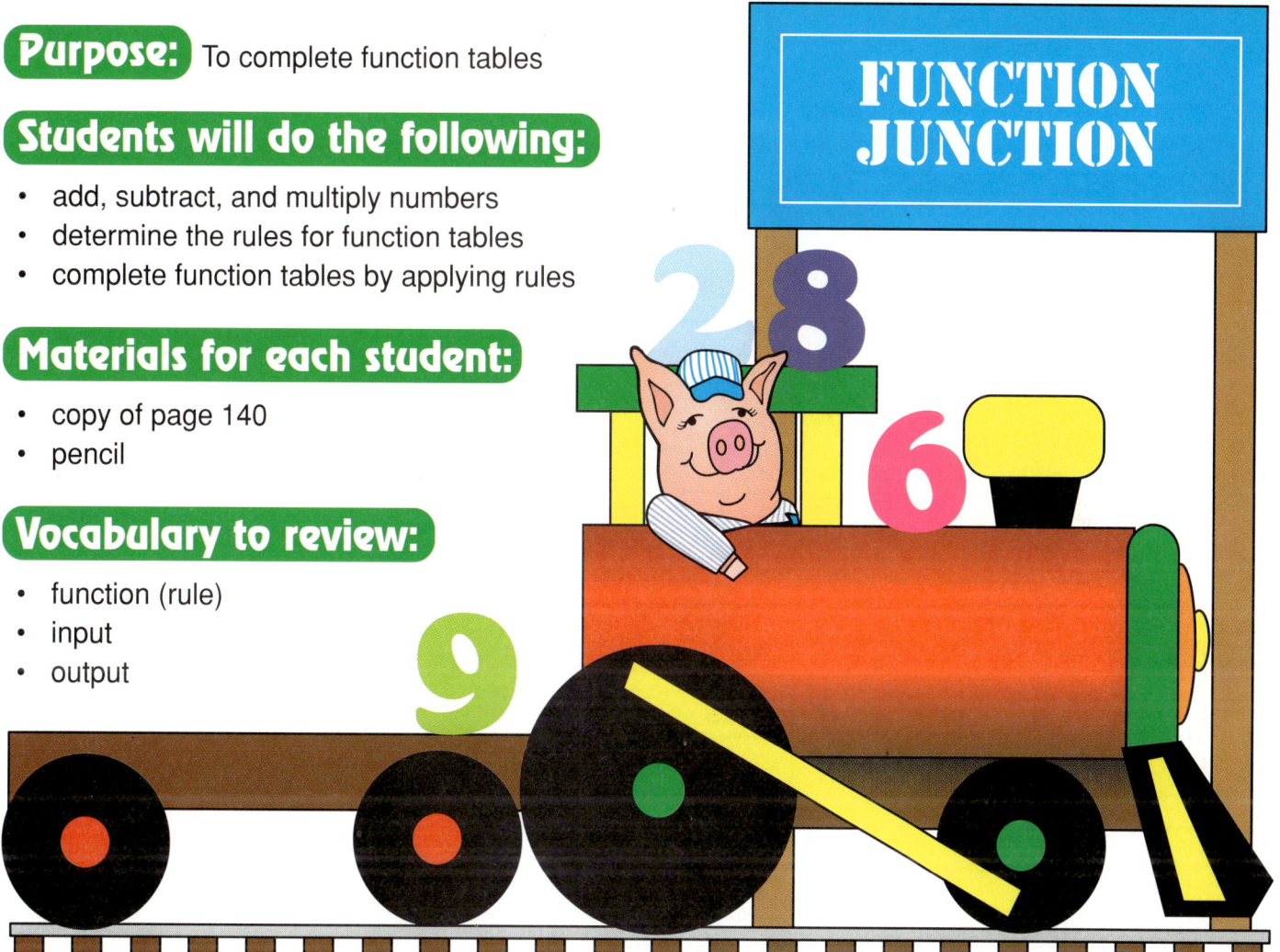

Purpose: To complete function tables

Students will do the following:
- add, subtract, and multiply numbers
- determine the rules for function tables
- complete function tables by applying rules

Materials for each student:
- copy of page 140
- pencil

Vocabulary to review:
- function (rule)
- input
- output

Extension activities to use after the reproducible:

- Have students communicate their understanding of function tables with this writing activity. Direct each student to choose one of the completed function tables from his copy of page 140. Have each student write to explain which table he chose and how he determined the function for the table. Encourage youngsters to draw examples to accompany their explanations as needed. Then provide time for each youngster to share his writing and strategy for completing the function table.

- Strengthen students' understanding of how to complete function tables with this activity. Instruct each student to draw a function table like the one shown. Then direct each student to randomly choose and write a number between 1 and 30 in each row of the input column. Next, have him randomly choose and write a number in the first row of the output column. Challenge each youngster to study the two numbers in the first row of his table and determine a function or rule for his table. Then ask students to use their rules to complete the output columns of their tables. If desired, allow time for students to trade papers and guess the function for other classmates' tables.

Input	Output

Function tables

Function Junction

Study each table to figure out its *function*, or rule. Then complete the table.
On each line, write the code for the function you used. Use the Function Code.

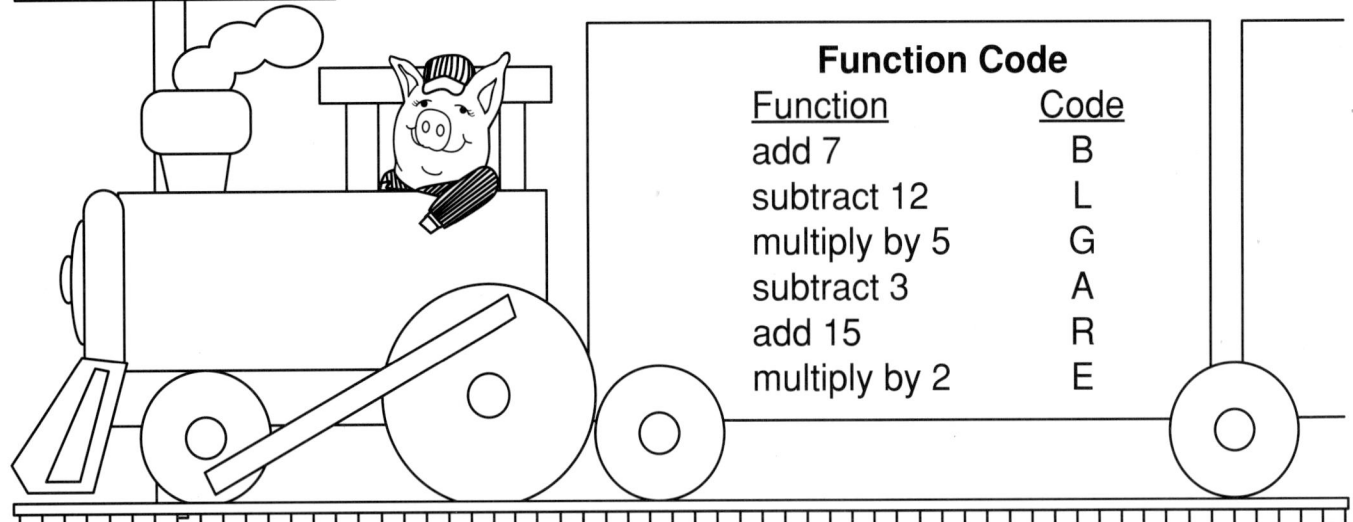

Function Code

Function	Code
add 7	B
subtract 12	L
multiply by 5	G
subtract 3	A
add 15	R
multiply by 2	E

1. Code: _____

Input	Output
10	7
14	11
6	
9	
	2

2. Code: _____

Input	Output
3	10
5	
14	
21	28
0	

3. Code: _____

Input	Output
3	15
	30
8	40
	10
10	

4. Code: _____

Input	Output
8	16
10	20
	8
	30
50	

5. Code: _____

Input	Output
5	20
15	30
20	
	45
10	

6. Code: _____

Input	Output
26	14
15	3
13	
	8
	0

What kind of math activity did you just complete?
To find out, match each letter to a numbered line below.

An ___ ___ ___ ___ ___ ___ ___ activity!
     1 6 3 4 2 5 1

Bonus Box: Make an input-output table of your own. Find a friend to figure out the function for your table.

Backward Birdie

Grow strong problem-solving skills as students fly through the working-backward strategy!

Purpose: To solve problems by working backward

Students will do the following:
- work backward by adding or subtracting
- calculate inverse operations

Materials for each student:
- copy of page 142
- pencil
- crayons

Vocabulary to review:
- working backward
- reverse

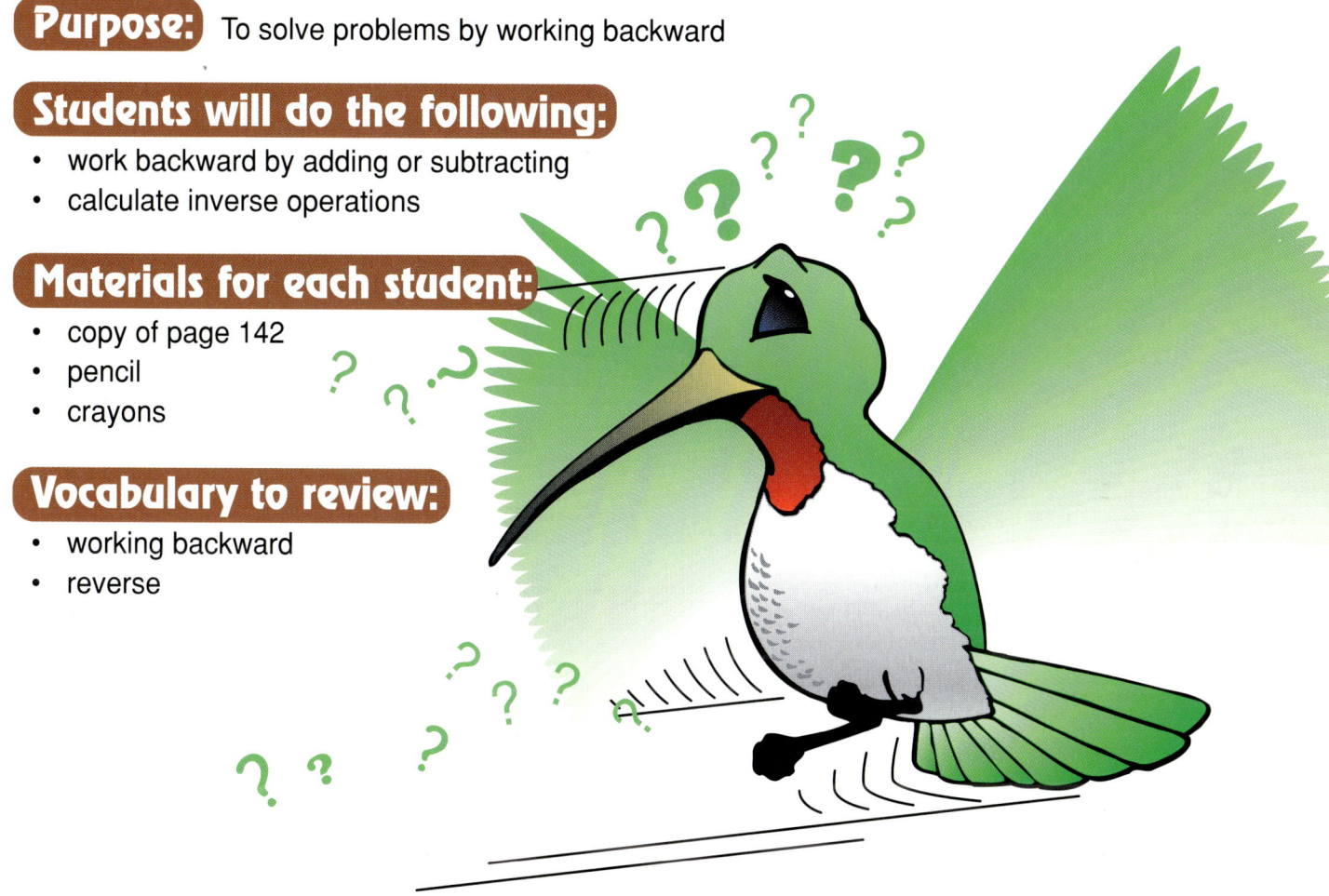

Extension activities to use after the reproducible:
- Watch students' problem-solving skills grow as they work backward to complete this daily measurement challenge. Prepare a large construction paper flower that measures 18 inches in length. Write on a large index card "Daily Growth Rate = 2 inches." Display the flower and card in the front of the room. Place a yardstick nearby. Write on the board "Today is [day of the week]. On what day was this flower 14 inches tall?" Invite students to solve the problem during their free time. To vary the problem each day, change the question, height of the flower, or growth rate.

- Provide independent practice with the working-backward strategy at this simple math center. Program each of 20 index cards, as shown, with a problem similar to the ones on page 142. Write the answer on the back of the card for self-checking. For each problem, program a corresponding clothespin with the answer. Place the cards and clothespins at a center. A student chooses a card and reads the problem. He works backward to solve it and then clips the clothespin with the matching answer to the card. He repeats this with the remaining cards and then checks his work by looking at the back.

Name _____

Work backward

Backward Birdie

Herbie Hummingbird does everything backward!
To SOLVE each problem, write it backward and reverse the addition or subtraction signs.
Then rewrite the problem to CHECK your answer.
If your answer works, color the matching flower.

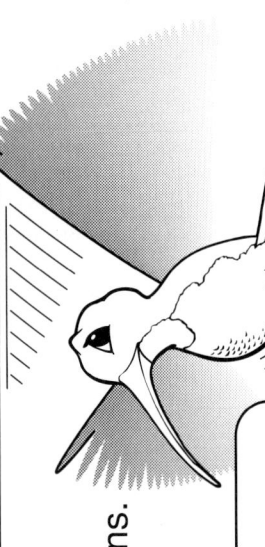

Example

Problem	Solve	Check
☐ + 3 − 1 = 12	12 + 1 − 3 = [10]	[10] + 3 − 1 = 12 ✓

A. ☐ − 2 + 6 = 12
B. ☐ − 9 + 3 = 9
C. ☐ − 6 + 1 = 7
D. ☐ + 5 − 7 = 8
E. ☐ − 1 + 8 = 14
F. ☐ + 1 − 10 = 0
G. ☐ − 8 + 5 = 13
H. ☐ + 4 − 2 = 8
I. ☐ − 4 + 6 = 15
J. ☐ − 8 + 9 = 18
K. ☐ + 8 + 3 = 16
L. ☐ − 2 + 7 = 10

Solve _____

Check _____

Bonus Box: Herbie had a party with some of his friends. Later that afternoon, 2 birds arrived and 5 went home. There were 6 birds left. How many birds were at the party when it began? (*Hint: Write a number sentence like the ones above.*)

Rows of Reasoning

Dig into this lesson to give youngsters practice using logical reasoning!

Purpose: To solve problems using logical reasoning

Students will do the following:
- interpret clues in word problems
- complete a grid using the process of elimination

Materials for each student:
- copy of page 144
- pencil
- crayons
- scissors
- glue

Vocabulary to review:
- logical reasoning

Extension activities to use after the reproducible:
- Line up logical-reasoning practice with this partner game. To play, one student in each pair draws three columns and labels the top of each one as shown. Then he writes a two-digit number on a separate slip of paper and conceals it from his partner. Next, the other student in the pair tries to guess the number. If he guesses a number that contains neither of the digits, his partner says "Zip" and writes the incorrect guess in the first column. If one digit is correct, his partner says "Zap" and writes the incorrect guess in the second column. If both of the digits are correct, but are in reverse order, his partner says "Zing" and writes the incorrect number in the last column. This continues until the student guesses the correct number. Then the partners switch roles and play again.

- Play a game of Secret Student to teach the process of elimination. In advance, secretly choose a student and write her name on a slip of paper. To begin the game, have the students stand; then invite them to ask yes or no questions about the secret student, such as the following: Is the secret student a girl? Does she ride the bus to school? Is she wearing a red shirt? After you respond to each question, instruct students who are eliminated to sit down. When one student is left standing, invite her to verify that she is the secret student by reading the slip of paper. Secretly choose another student for the next round of play.

Logical reasoning

Name _____ Logical reasoning

Rows of Reasoning

Katie forgot to mark the plants in her garden!
Color and cut out each plant marker.
Use the clues to place each marker.
Then glue the markers in place.

Row 1
Row 2
Row 3
Row 4

Clues:
- Katie planted 2 tomato plants in Row 1.
- There are no carrots in Row 1 or Row 2.
- Each corner has a pepper plant.
- The pea plants are all in the same row.
- There are no tomato plants in Row 3.

Bonus Box: Explain how the third clue above helped you decide where to put the tomato plants.

©2001 The Education Center, Inc. • Math Skills Workout • TEC3227 • Key p. 175

144

Tricks of the Trade

Have youngsters act out problems with trading cards!

Purpose: To solve problems by acting them out

Students will do the following:
- interpret word problems
- use manipulatives to problem solve
- act out solutions to word problems

Materials for each student:
- copy of page 146
- pencil
- scissors

Vocabulary to review:
- trade

Extension activities to use after the reproducible:

- Extend the activity on page 146 to give youngsters more practice acting out problems. Draw a chart on the chalkboard like the one shown. Provide each student with a trading card, a piece of candy, or some other small item to trade. Then divide students into pairs. Ask each student to trade her item with her partner. On the chart record that two children made one trade. Divide students into groups of three. Have students act out how many different trades can be made in the group if each person trades with the other one time. Record the correct number of trades on the chart (*three trades*). Continue this process, dividing students into groups of four, then five (*six trades, ten trades*). Have students identify a pattern shown in the results on the chart. Then have students predict how many trades can be made with six students (*15 trades*). Divide students into groups of six to verify their predictions.

Number of Students	2	3	4	5	6
Number of Trades					

- Put a little rhythm in your problem-solving practice. Slowly clap an even rhythm and invite students to snap on every other clap. Then have students predict and act out how many times they will snap if there are 21 claps. Repeat this activity by clapping a different number of times and changing the snapping frequency to every third or fourth clap.

Act it out

Name_____ Act it out

Tricks of the Trade

Cut out the trading cards below.
Give the baseball cards to Lisa, the football cards to
   Freddy, and the basketball cards to Brian.
Then use the cards to solve each problem.
Write your answer in the matching box.

1. On Sunday, Lisa trades 2 baseball cards for 3 of Freddy's football cards. How many cards does Lisa have?
2. On Monday, Brian trades 2 basketball cards for 1 of Lisa's baseball cards and 1 of her football cards. How many football cards does Lisa have?
3. On Tuesday, Freddy trades 1 baseball card for 2 of Brian's basketball cards. Who has the fewest cards?
4. On Wednesday, Brian trades 1 basketball card for 2 of Lisa's football cards. How many football cards does Brian have?
5. On Thursday, Freddy trades all of his basketball cards for all of Brian's football cards. How many basketball cards does Brian have?
6. On Friday, Lisa trades 2 baseball cards for 1 of Freddy's football cards. Who has the most sports cards?
7. On Saturday, Freddy gives 1 baseball card to each of his friends. How many sports cards does Freddy have now?

1.	2.	3.	4.	5.	6.	7.

Bonus Box: Was it helpful to use the trading cards to solve each problem? Write to explain why or why not.

©2001 The Education Center, Inc. • *Math Skills Workout* • TEC3227 • Key p. 175

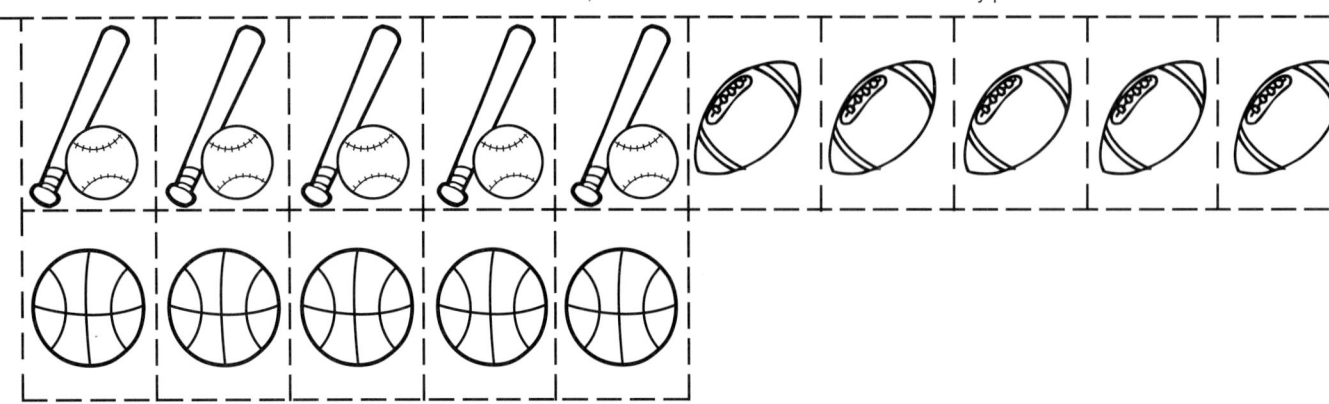

Flower Power

Watch students' thinking skills blossom as they draw pictures to solve problems!

Purpose: To solve problems by drawing pictures

Students will do the following:
- follow directions to draw a Hawaiian lei
- make and use drawings to solve problems
- add, subtract, and compare numbers

Materials for each student:
- copy of page 148
- pencil
- crayons

Vocabulary to review:
- lei
- more
- fewer

Extension activities to use after the reproducible:

- Involve students in creating daily problem-solving-practice activities. Pair students and challenge each pair to write a problem that can be solved by drawing a picture. Have each pair solve the problem on the back of its sheet. Duplicate each pair's problem onto a transparency. When it's time for problem-solving practice, display one of the transparencies for the class. Upon completion of the problem, invite the student pair that created the problem to explain and model its solution.

- Students will love these personalized problem-solving challenges. Create a list of problems similar to the ones shown that include the names of the students in your class. Design enough problems so that each student's name is included once. Provide each student with a copy of the list. Read the first problem aloud and have students solve it by drawing a picture. Next, invite the students whose names are included in the problem to come to the front of the room and act out the problem to display the correct answer. Repeat this process with each of the remaining problems on the list.

1. Todd, Doreen, Merita, Jay, and Al line up. Al stands behind Merita and in front of Doreen. Doreen stands in front of Todd. No one is standing in front of Jay. In what order are the students lined up?

2. Myra, Kenny, Ericka, Mark, Jamie, and Nikki line up. Myra is first in line. Kenny is between Ericka and Myra. Mark is between Jamie and Nikki. Jamie is not beside Ericka.

3. Mac, Tyler, Nicole, Laurel, Jeremy, and Bryan are standing in two lines of three. Mac is to the right of Nicole. Tyler is behind Mac but in front of Jeremy. Bryan is last in his line. In what order are the students lined up?

4. Six students are sitting in a circle. David and Derrick are across from each other. Angie is next to Derrick. Stan is between Angie and David. Tanya and Ashley are beside each other. Ashley is between Derrick and Tanya. In what order are the students in the circle?

Draw a picture

Name_____   *Draw a picture*

Flower Power

Color the flowers on Patty's lei blue.

Read the clues.
Draw and color more flowers to complete Patty's lei.

Clues
- There are 2 more yellow flowers than blue.
- There are 3 fewer red flowers than yellow.
- There are twice as many orange flowers as red.
- There are 4 fewer purple flowers than yellow.

1. What color flower is most often used? _____
2. How many orange and blue flowers are there all together? _____
3. How many more yellow than purple flowers are there? _____
4. How many flowers are not blue or orange? _____
5. What is the total number of flowers on Patty's lei? _____

Bonus Box: If a lei has 9 flowers and there are 5 more red than orange flowers, how many of the flowers are red?

Ready to Race!

Keep youngsters on track when it comes to solving problems by making tables!

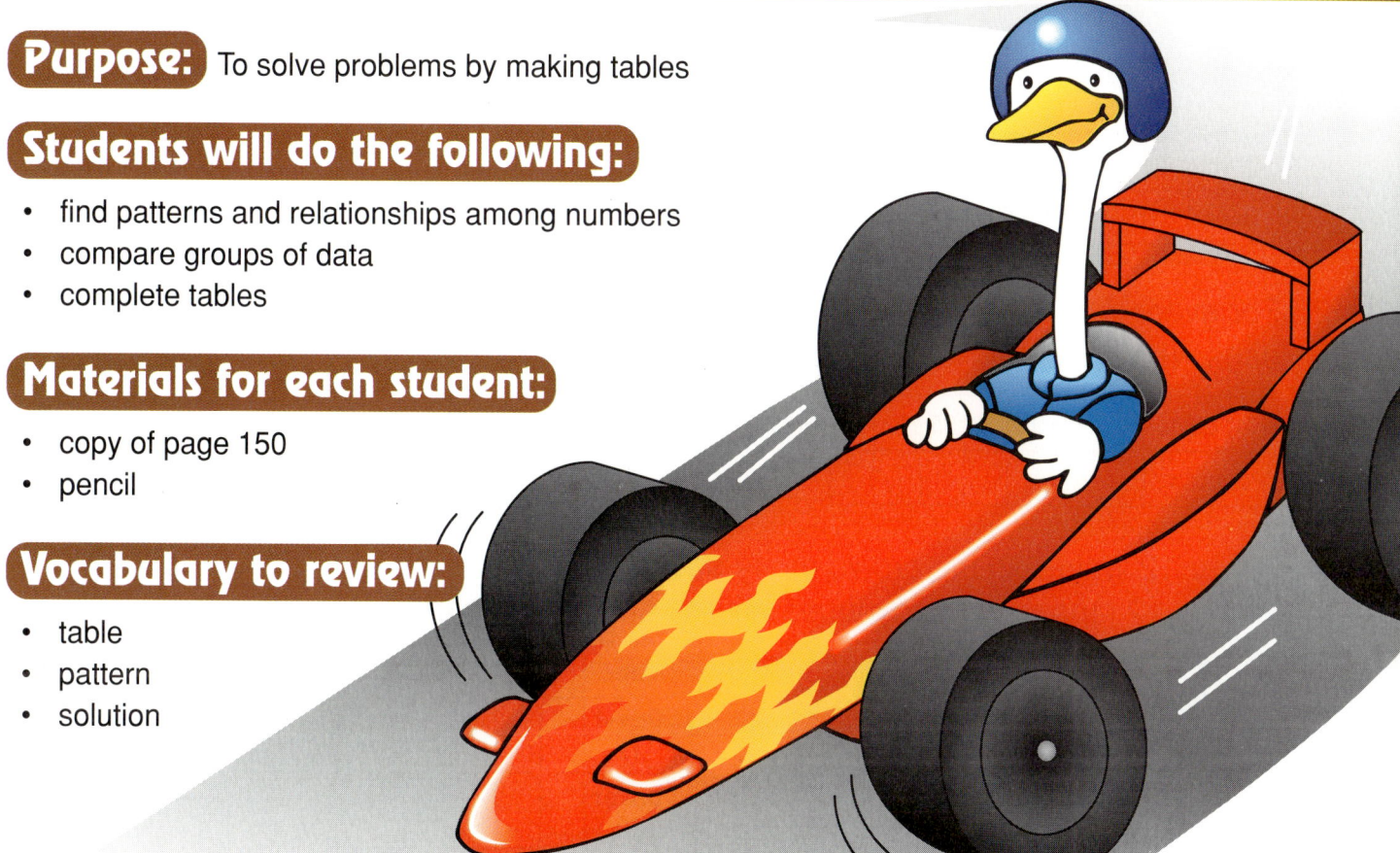

Purpose: To solve problems by making tables

Students will do the following:
- find patterns and relationships among numbers
- compare groups of data
- complete tables

Materials for each student:
- copy of page 150
- pencil

Vocabulary to review:
- table
- pattern
- solution

Extension activities to use after the reproducible:

- Hold a little race of your own to reinforce the make-a-table strategy! At a predetermined signal, direct each student to run in place for ten seconds, counting each step. Next, have each student draw a table that shows the relationship between increments of ten seconds and the number of steps she took in that time. Then have her complete the table to determine how many steps she could take in one minute. Repeat this activity in a similar manner, having students display a different physical activity, such as jumping jacks or hopping on one foot, for different increments of time.

- For additional practice using the make-a-table strategy, draw a table like the one shown. Pair students and have each pair complete and extend the table to solve this challenging problem: *During the school week, Jane washes the chalkboard every morning and afternoon, Chris empties the pencil sharpener every afternoon, and Tina collects journals every other afternoon. If this pattern continues, how often will all three classmates work on their tasks at the same time over a two-week period?* Although each twosome's table may be different—depending on if they show Tina's first afternoon on Monday or Tuesday—they should determine that the students work on their tasks simultaneously five times within a two-week period.

Day	Monday		Tuesday		Wednesday	
	A.M.	P.M.	A.M.	P.M.	A.M.	P.M.
Jane						
Chris						
Tina						

Make a table

Name_____ Make a table

Ready to Race!

Read each problem.
Complete each table to help you solve each problem.
Then write each solution in a complete sentence.

1. Each of 12 cars needed all 4 tires changed. The pit crews only had 40 tires. How many cars did not get tires?

Cars	1	2	3	4	5	6	7	8	9	10	11	12
Tires												

Solution: _____

2. For every 7 hot dogs sold, hungry fans bought 3 bags of peanuts. How many bags were sold after 49 hot dogs were sold?

Hot Dogs	7	14	21	28	35	42	49
Peanuts							

Solution: _____

3. Driver 12 waved to the crowd every second lap. Driver 30 waved every third lap. In 12 laps, how many times did the drivers wave in the same lap?

Solution: _____

Laps	1	2	3	4	5	6	7	8	9	10	11	12
Driver 12												
Driver 30												

4. After 2 laps, 2 cars got gas. After the next 2 laps, 3 more cars got gas. After the third 2 laps, 4 more cars got gas. If this pattern continued, how many got gas after 12 laps?

Solution: _____

Laps	2	4	6	8	10	12
Cars						

5. Fifteen fans were allowed to meet the drivers. Every fourth fan was a child. How many children met the drivers?

Fans	1	2	3	4	5	6	7	8	9	10	11	12	13	14	15
Adult															
Child															

Solution: _____

Bonus Box: If the number of fans in problem 5 were 23, how many fans would be children? Extend the table on a separate sheet of paper to find the answer.

Score With Patterns!

Watch students get a kick out of solving these problems by finding patterns!

Purpose:
To solve problems by identifying patterns

Students will do the following:
- identify number patterns and their rules
- read a calendar to find a pattern
- continue patterns to find solutions

Materials for each student:
- copy of page 152
- pencil

Vocabulary to review:
- pattern
- rule

Extension activities to use after the reproducible:

- Students will enjoy constructing patterns with this problem-solving task. Provide each student with 25 interlocking cubes. Lead students to build the three figures shown in Example 1 one at a time. Each time, have students identify the number of cubes needed to build each figure; then record the number on the chalkboard. Have students study the figures and numbers displayed and describe the pattern shown. *(Two cubes are added each time.)* Direct students to build the fourth figure in the pattern. Then challenge youngsters to use their cubes to answer questions such as the following: How many blocks are needed to build the eighth figure? *(15)* If you continue the pattern, what is the total number of figures you can build with 25 cubes? *(13)* How many more cubes would you need to build the 14th figure in the pattern? *(2)* Then repeat this activity, having students build the figures shown in Example 2.

- Use this challenging partner game to give youngsters more practice solving problems with patterns. Post the scoring chart shown. Then provide each pair of students with a paper bag containing 30 interlocking cubes (six of each color shown on the scoring chart). To play, students in each pair take turns drawing a cube from the bag until each player has a total of 15 cubes. Then each player strategically thinks about how to earn the higher point value as he tries to use at least 10 of his cubes to form a color pattern. If the player is able to use at least 10 cubes, he uses the chart to total his score. The player with more points wins the round.

Example 1

Figure 1   Figure 2   Figure 3

Example 2

Figure 1   Figure 2   Figure 3

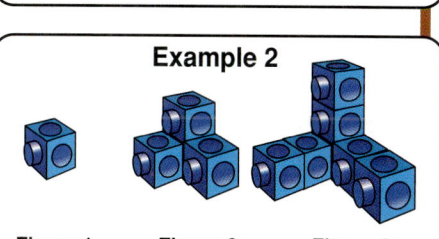

Scoring Chart
red = 5 points
yellow = 4 points
blue = 3 points
green = 2 points
orange = 1 point

Find a pattern

Name _____ Find a pattern

Score With Patterns!

Help Rita Rabbit complete her soccer schedule for the month. Use the calendar and key to find patterns in her schedule. Then draw the correct symbols on the calendar to show Rita's
- next game
- next scrimmage
- next 2 practices

KEY
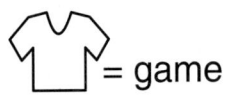 = game  = scrimmage 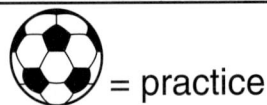 = practice

October

Sunday	Monday	Tuesday	Wednesday	Thursday	Friday	Saturday
	1 👕	2	3	4 🚧	5	6 ⚽
7	8 🚧	9 👕	10	11 ⚽	12	13
14	15	16 ⚽	17 👕	18 🚧	19	20
21 ⚽	22	23	24	25	26	27
28	29	30	31			

Complete each sentence.

1. The dates for Rita's practices are October 6, ____, ____, ____, ____, and ____. The rule for the pattern is _____.
2. The dates for Rita's scrimmages are October 4, ____, ____, and ____. The rule for the pattern is _____.
3. The dates for Rita's games are October 1, ____, ____, and ____. The rule for the pattern is _____.
4. The day of the week for Rita's next game will be _____.
5. Next month, the date for Rita's first game will be _____, the date for her first scrimmage will be _____, and the date for her first practice will be _____.

Bonus Box: If the coach adds a practice on October 5 and then has the team meet every 5 days to practice again, on which date will Rita have 2 soccer activities scheduled?

152 ©2001 The Education Center, Inc. • *Math Skills Workout* • TEC3227 • Key p. 176

Mall Mania

Sell students on using the make-a-list strategy to solve problems!

Purpose: To solve problems by making lists

Students will do the following:
- organize groups of items into lists
- list all possible combinations of a group of items
- use reasoning skills

Materials for each student:
- copy of page 154
- pencil

Vocabulary to review:
- combination
- possible

Extension activities to use after the reproducible:

- Tickle students' taste buds with this problem-solving task! Challenge each student to list all the possible combinations of ice-cream sundaes—containing one flavor, one syrup, and one topping—that can be made from two different ice-cream flavors, two different ice-cream syrups, and two different toppings *(eight combinations)*. Next, have each student draw and label a picture to show his favorite combination. If desired, top off this activity by serving each student his favorite sundae combination as a treat!

- Have students explore place value using the make-a-list strategy. Supply each student with a set of number cards (pattern on page 166). Select and state three number cards for students to use, such as 1, 2, and 3. Then challenge students with a problem, such as finding all the possible combinations with 2 in the ones place. After the possible combinations have been determined, have students select four number cards to repeat a similar challenge. To increase difficulty, have students increase the number of cards for the challenge, and designate which numbers should be in both the ones and thousands places.

Name _____ Make a list

Mall Mania

Liza is visiting the mall.
She has lots of decisions to make.
Make a list of Liza's choices in each matching box.
Then, in each circle, write her total number of choices.

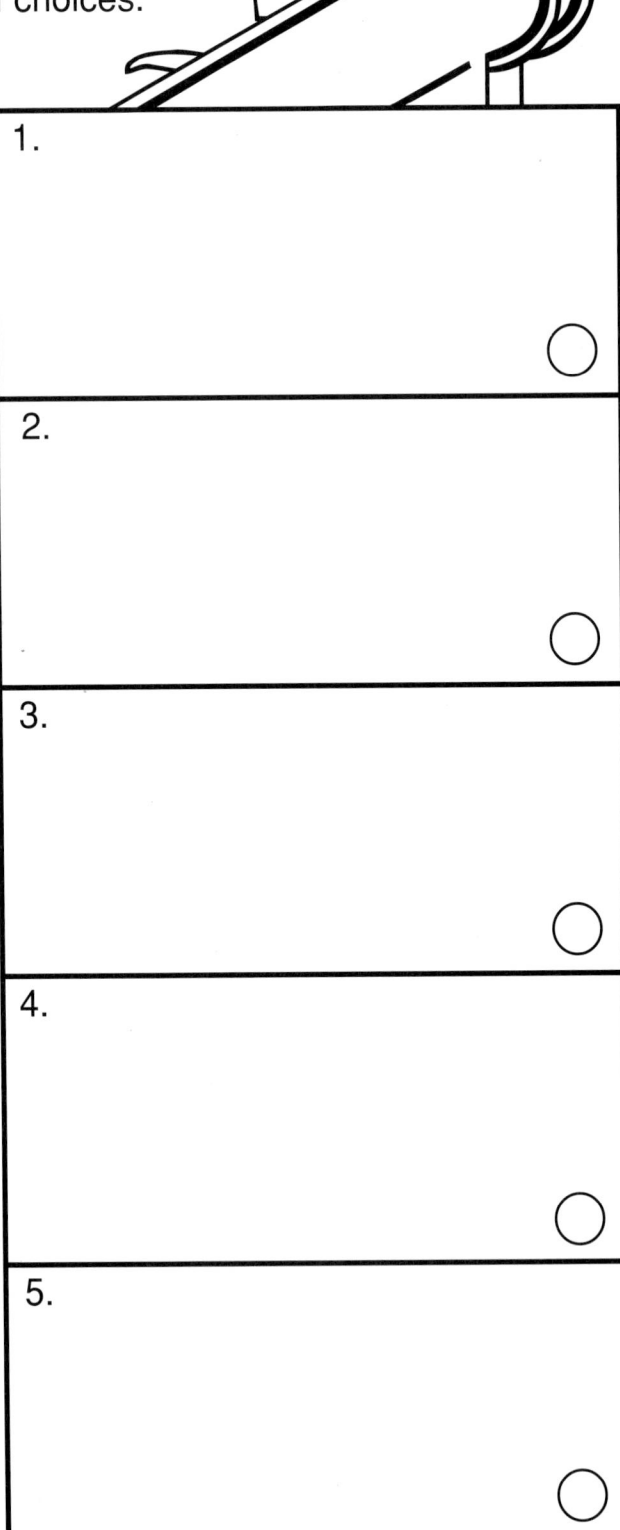

1. Liza's first stop is Moe's Music. She can choose a CD or cassette from either the Side Street Boyz, N'Line, or Jennifer Mopeds. List all of her possible choices.

2. Next, at Jazzy Jewelry, Liza can choose from a 16- or 18-inch gold or silver chain. She can add a heart, a star, or a flower charm. List her possible combinations.

3. At Fancy Feet, Liza can choose a pair of clogs, sandals, or sneakers. The colors available are red, blue, and yellow. List the different pairs of shoes Liza can buy.

4. At Clothes Closet, Liza shops for an outfit. She must decide between an orange or green shirt and tan or black pants. List the different outfit combinations she could buy.

5. At Candy Corner, Liza can get two fudge flavors for the price of one. She can choose between chocolate, vanilla, peanut butter, or maple. List her possible combinations.

Bonus Box: Write another problem about Liza at the mall. List the possible combinations.

Carnival Capers

Put students' problem-solving skills to the test as they engage in these carnival capers!

Purpose: To solve problems using the guess-and-check strategy

Students will do the following:
- use number sense to make reasonable guesses
- test guesses by using addition and subtraction

Materials for each student:
- copy of page 156
- pencil
- scratch paper

Vocabulary to review:
- guess and check
- reasonable

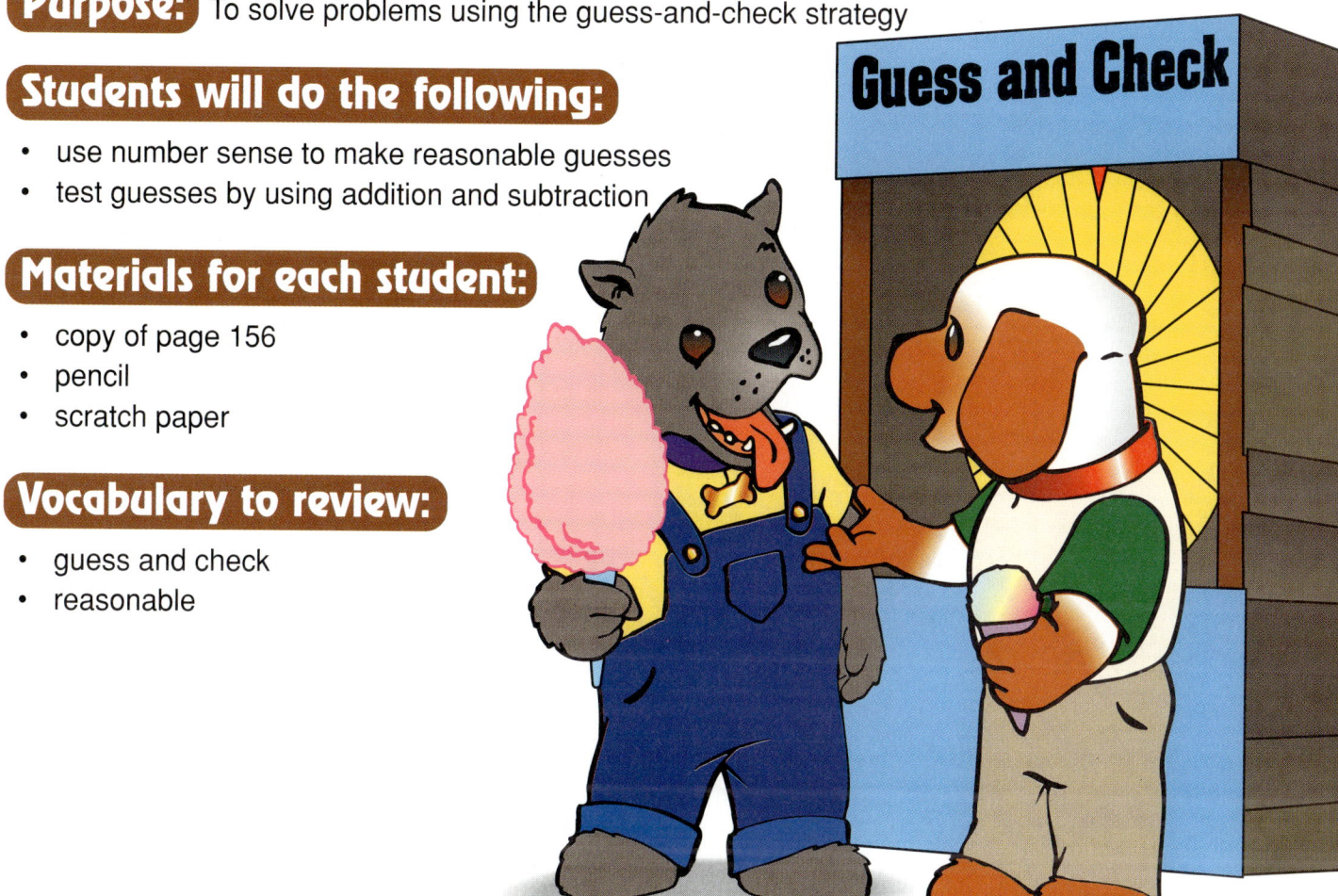

Extension activities to use after the reproducible:

- Keep the problem-solving practice rolling with this simple partner game. Give each pair a number cube. Instruct each partner to secretly roll the cube four times and add the four digits together. Have the students in each pair reveal their sums to each other. Then have the students in each pair take turns trying to guess and check the four numbers the other partner rolled to get his sum. Vary the activity by changing the number of times the cube is rolled.

- Bring out the dominoes to put your youngsters' problem-solving talents to work! Provide each student with a set of dominoes or cutout construction paper domino manipulatives (pattern on page 163). Challenge each youngster to model a magic square in which the dots on each side add up to six. (See the example shown.) Encourage youngsters to guess and check a variety of possibilities. Finally, have each student illustrate her magic square on a sheet of drawing paper. For additional domino challenges, have students create magic squares in which the sum of the sides is 9, 12, or 15.

Guess and check

Name _____

Guess and check

Carnival Capers

Read the problems to learn about Sam and Juan's day at the carnival. Use the guess-and-check strategy to solve the problems in order.

Tickets
Ride.....$2.50   Game.....$2.00
Food.....$1.00

1. Sam and Juan each spent $35.00 on ride, game, and food tickets. Sam bought 18 tickets and Juan bought 17 tickets. What combination of tickets could each boy have bought?

Sam: _____
Juan: _____

Pins to Win!

| 2 | 4 | 6 | 8 | 10 | 12 |

2. Sam and Juan used game tickets to play Pins to Win! Sam hit 4 pins to make 28 points. Juan hit 4 pins to make 30 points. Which pins did each player hit?

Sam: _____
Juan: _____

Toys!
Jacks	Glow Ball	Beanbag
$0.56	$0.67	$0.39
Jump Rope	Yo-Yo	Whistle
$0.77	$0.65	$0.19

4. Sam and Juan each spent 1 game ticket to buy toys. Sam bought 3 toys and Juan bought 4 toys. Which toys did they each buy?

Sam: _____
Juan: _____

Concession Stand
Burger	Salad	Soda
$2.40	$1.35	$1.00
Hot Dog	Nachos	Water
$2.00	$1.60	$0.65

5. Sam bought 3 items with 5 food tickets and Juan bought 3 items with 4 food tickets. What did they each buy to eat?

Sam: _____
Juan: _____

Wham-o! Bumper Cars

3. Juan rode 4 more times than Sam. If they rode a total of 12 times, how many times did they each go on the ride?

Sam: _____   Juan: _____

Bonus Box: Problem 2 has extra information not needed to solve the problem. Reread the problem and underline the extra information.

The Race Is On!

Help your youngsters get off to a good start by helping them choose the correct operation when solving problems!

Purpose: To solve problems by choosing the correct operation

Students will do the following:
- choose the correct operation for solving problems
- solve word problems
- add, subtract, multiply, and divide

Materials for each student:
- copy of page 158
- pencil
- crayons

Vocabulary to review:
- addition clue words: in all, total
- subtraction clue words: more than, many left, less than
- multiplication clue words: in all, all together, each
- division clue words: each, equally sharing

Extension activities to use after the reproducible:

- Strengthen students' problem-solving skills with this brain-building exercise! Challenge student pairs to solve the following problems by identifying the missing addition and subtraction signs. Write each problem on the chalkboard for students to copy. Then have each pair correctly place the operation signs in each problem.

A. 3 ◯ 2 ◯ 1 ◯ 4 ◯ 1 ◯ 3 = 10	(+, –, +, –, +)
B. 7 ◯ 4 ◯ 9 ◯ 5 ◯ 7 ◯ 2 = 8	(–, +, +, –, –)
C. 9 ◯ 2 ◯ 7 ◯ 8 ◯ 1 ◯ 9 = 0	(–, –, +, +, –)
D. 3 ◯ 3 ◯ 4 ◯ 4 ◯ 5 ◯ 5 = 4	(+, +, +, –, –)
E. 2 ◯ 2 ◯ 9 ◯ 7 ◯ 5 ◯ 4 = 15	(–, +, +, –, +)

- Student's problem-solving talents are sure to shine with this classroom competition. Draw a large star on the chalkboard, and write a number in the star, such as 100. Then divide students into teams of three or four. Challenge each team to write a problem that reaches the star number using two operations (for example: 75 + 50 – 25 = 100). The first team to meet the challenge earns a point for the round. Repeat this process until one team wins the game by earning three points. For a more challenging game, direct teams to reach the target number with specific combinations of operations, such as add/subtract or subtract/multiply.

Choose the operation

Name _____ Choose the operation

The Race Is On!

Read each problem about the Racers Athletic Club (RAC) runners.
Choose the best operation to solve each problem.
Use the code to color each ribbon.
Then solve each problem.

Color Code
addition = red   multiplication = yellow
subtraction = blue   division = green

 1. The RAC entered 5 teams in the 400-meter relay. If there were 4 runners on each team, what was the total number of RAC runners in the race?	2. At practice, the team warmed up for 15 minutes, ran for an hour, and cooled down for 15 minutes. What was the total time spent at practice?
3. At the meet, 4 runners run a total of 12 races. If they each run the same number of events, how many races will each run?	4. There are 16 hurdle racers and 20 sprinters on the team. How many more sprinters are there than racers?
5. At practice, Rita Rabbit ran 8 miles a day for 5 days. What was her total number of miles?	6. Harry Hare ran 8 miles less than Rita. How many miles did he run?
7. The team won 14 races at the first meet, 17 races at the second meet, and 12 races at the third meet. How many races did they win in all?	8. The RAC is scheduled for 30 meets this year. The members have already run in 7 meets. How many more meets do they have left?
9. The RAC team has 36 runners. For practice the team is put into 6 equal squads. How many runners are on each squad?	10. The RAC ordered 2 uniforms for each of the 36 runners. How many uniforms did they order in all?

Bonus Box: Tell how the answer would change for problem 3 if 6 runners ran a total of 12 races.

Big Top Strategies

Bring problem-solving strategies into the center ring!

Purpose: To solve problems by choosing strategies

Students will do the following:
- interpret word problems
- choose the best strategy to solve a problem

Materials for each student:
- copy of page 160
- pencil
- scissors
- glue

Vocabulary to review:
- make a list
- find a pattern
- act it out
- make a table
- use logical reasoning
- guess and check
- draw a picture
- choose the operation
- work backward

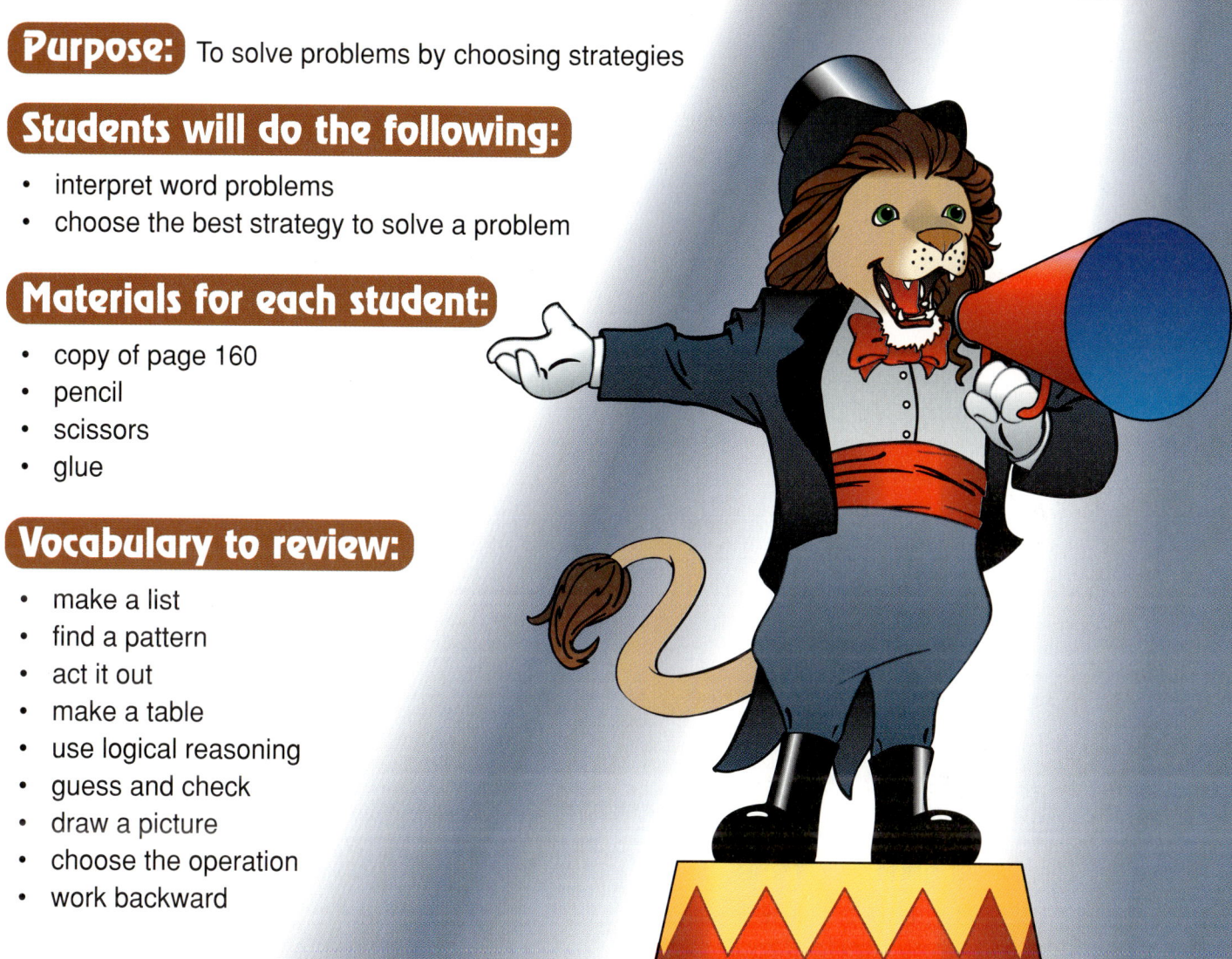

Extension activities to use after the reproducible:
- Begin every day the problem-solving way! Each morning post a problem for students to solve. Challenge each student to solve the problem, identify the strategy used, and deposit his folded paper into a box. After students are finished, randomly draw a predetermined number of papers. Give a small treat to each student who listed an appropriate strategy and has the correct solution.

- Use this handy reference book to put students over the top with problem solving! Provide each student with five sheets of unlined paper. Direct each student to fold her stack in half, staple it along the fold, and title her booklet as shown. Then have her label each booklet page with each of the strategies listed in "Vocabulary to review" on this page. As your class studies each strategy, have students define the strategy on the corresponding page and write a sample problem that is appropriate for the strategy. Upon completion of the books, encourage students to use them to help them choose the correct strategy in future problem-solving situations.

Choose a strategy

Name _____   Choose a strategy

Big Top Strategies

Cut out the problem-solving strategies.
Read the problems and then choose the best
strategy to solve each problem.
Solve the problem to test each strategy.
If the strategy works, glue it in place.

1. There are 4 animals in the parade. The bear is in front of the tiger. The horse is behind the tiger. The bear is behind the elephant. Which animal leads the parade?

2. The trapeze artists—Jo, Mo, Flo, and Bo—swing 2 at a time. How many combinations of artists can swing together? _____

3. The monkey and tiger were in the ring for a total of 34 minutes. The monkey was in the ring 10 minutes longer than the tiger. How long was each animal in the ring?
monkey _____
tiger _____

4. The vendors sell 3 boxes of popcorn for $5.00. How many boxes of popcorn can you buy for $25.00? _____

5. The acrobats made a pyramid. There was 1 acrobat on top, 3 in the next row, and 5 in the next row. If they made 5 rows, how many acrobats were on the bottom? _____

6. Kim, Tim, and Jim each had a snack at the circus: a candied apple, popcorn, or cotton candy. Kim dislikes sweets. Jim prefers fruit. Which snack did each friend eat?
Kim _____
Tim _____
Jim _____

Bonus Box: Three strategies are missing from this page—*draw a picture, work backward,* and *choose an operation.* Use one of these strategies to solve problem 6 again.

©2001 The Education Center, Inc. • *Math Skills Workout* • TEC3227 • Key p. 176

| make a list | find a pattern | act it out | make a table | use logical reasoning | guess and check |

Hundreds Chart

1	2	3	4	5	6	7	8	9	10
11	12	13	14	15	16	17	18	19	20
21	22	23	24	25	26	27	28	29	30
31	32	33	34	35	36	37	38	39	40
41	42	43	44	45	46	47	48	49	50
51	52	53	54	55	56	57	58	59	60
61	62	63	64	65	66	67	68	69	70
71	72	73	74	75	76	77	78	79	80
81	82	83	84	85	86	87	88	89	90
91	92	93	94	95	96	97	98	99	100

©2001 The Education Center, Inc. • *Math Skills Workout* • TEC3227

Cube

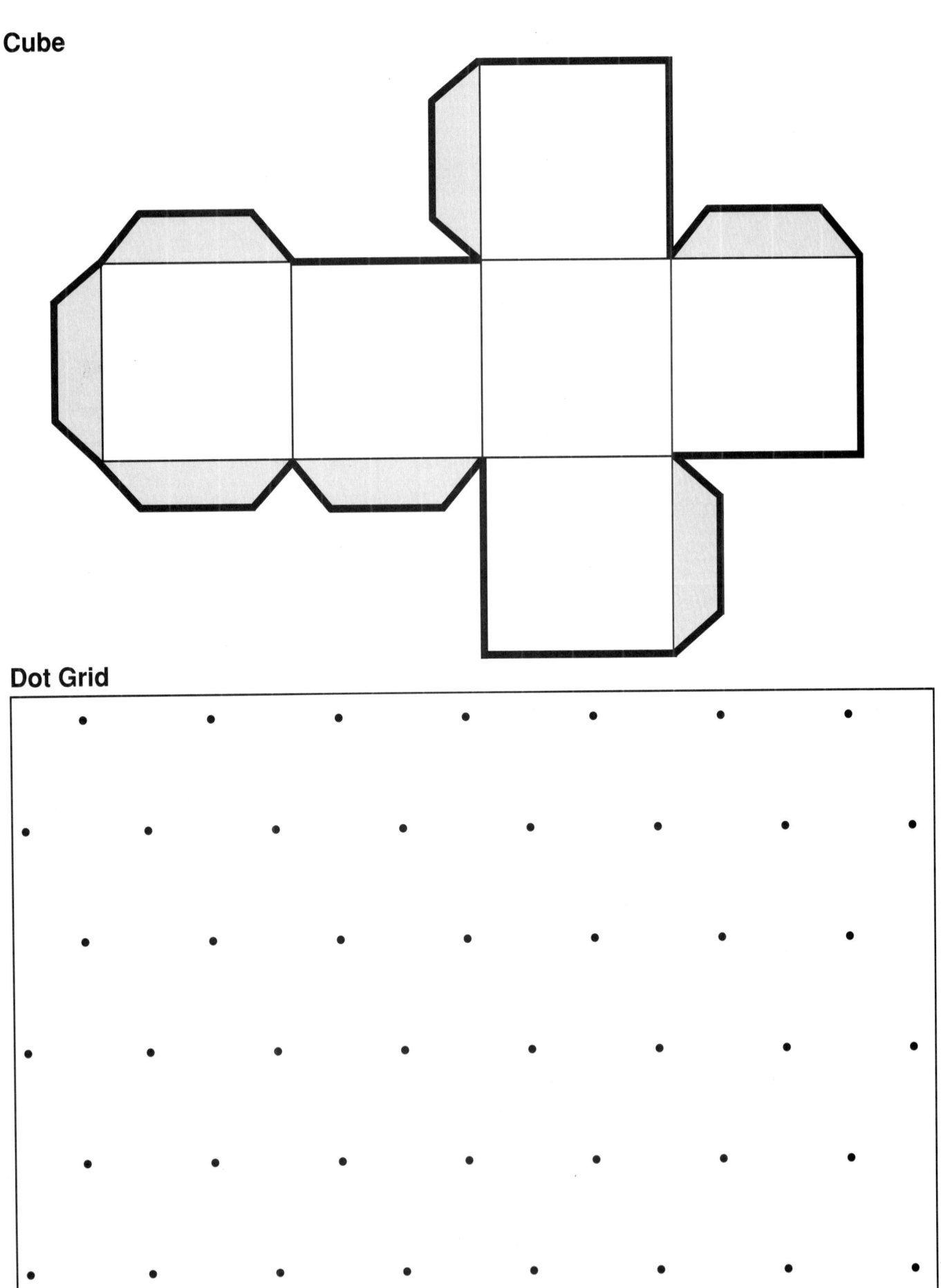

Dot Grid

©2001 The Education Center, Inc. • Math Skills Workout • TEC3227

Note to the teacher: To make a cube or die, make a construction paper copy of the cube pattern, program it as desired, and then cut it out along the bold lines. Next, fold the pattern along the thin lines and shape it into a cube. Then glue or tape each tab inside its adjacent face.

Dominoes

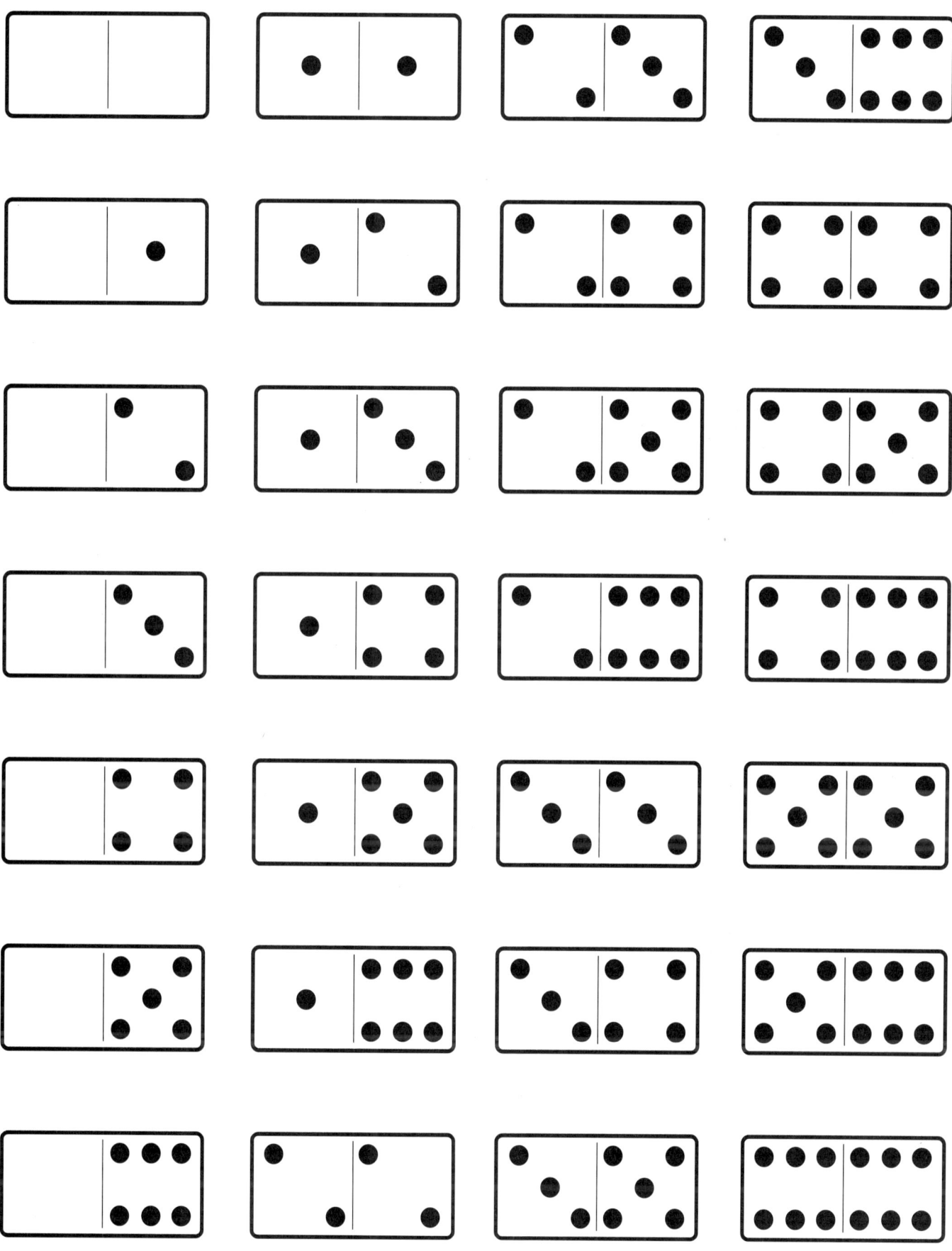

Clock and Hands

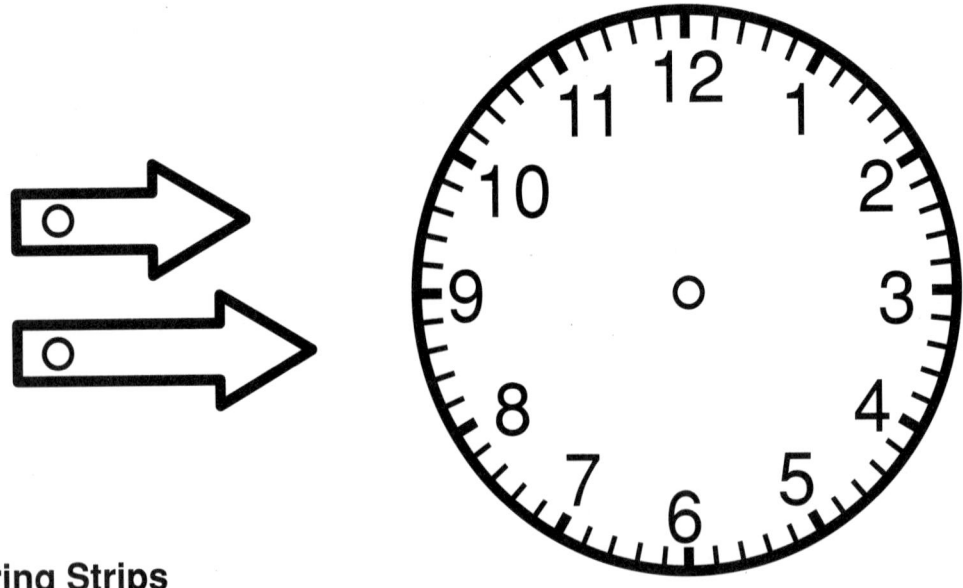

Measuring Strips

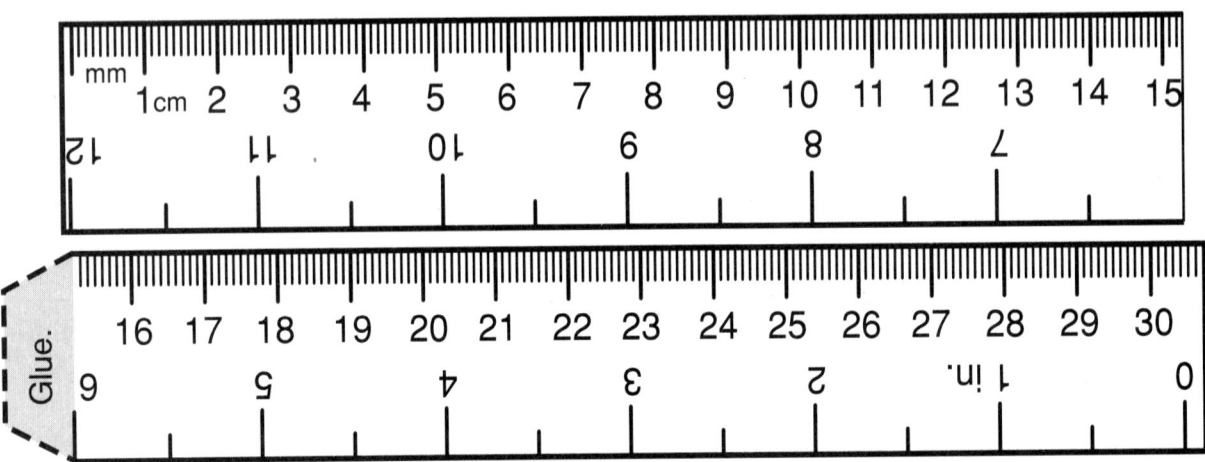

Thermometer

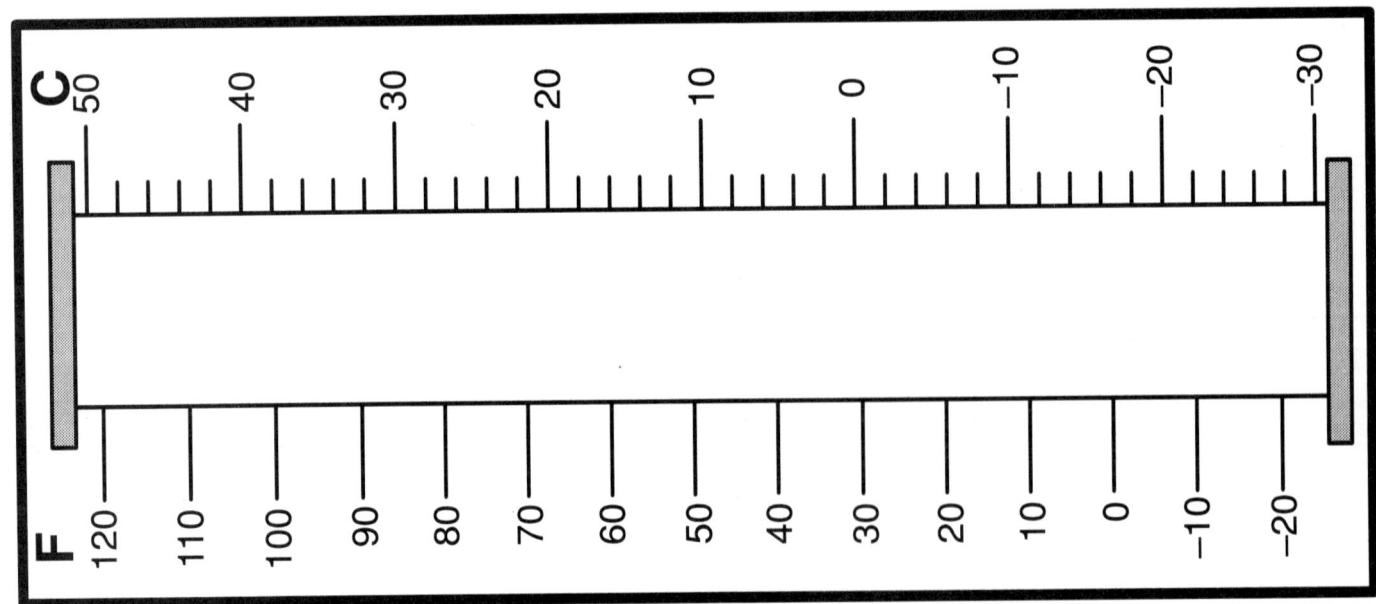

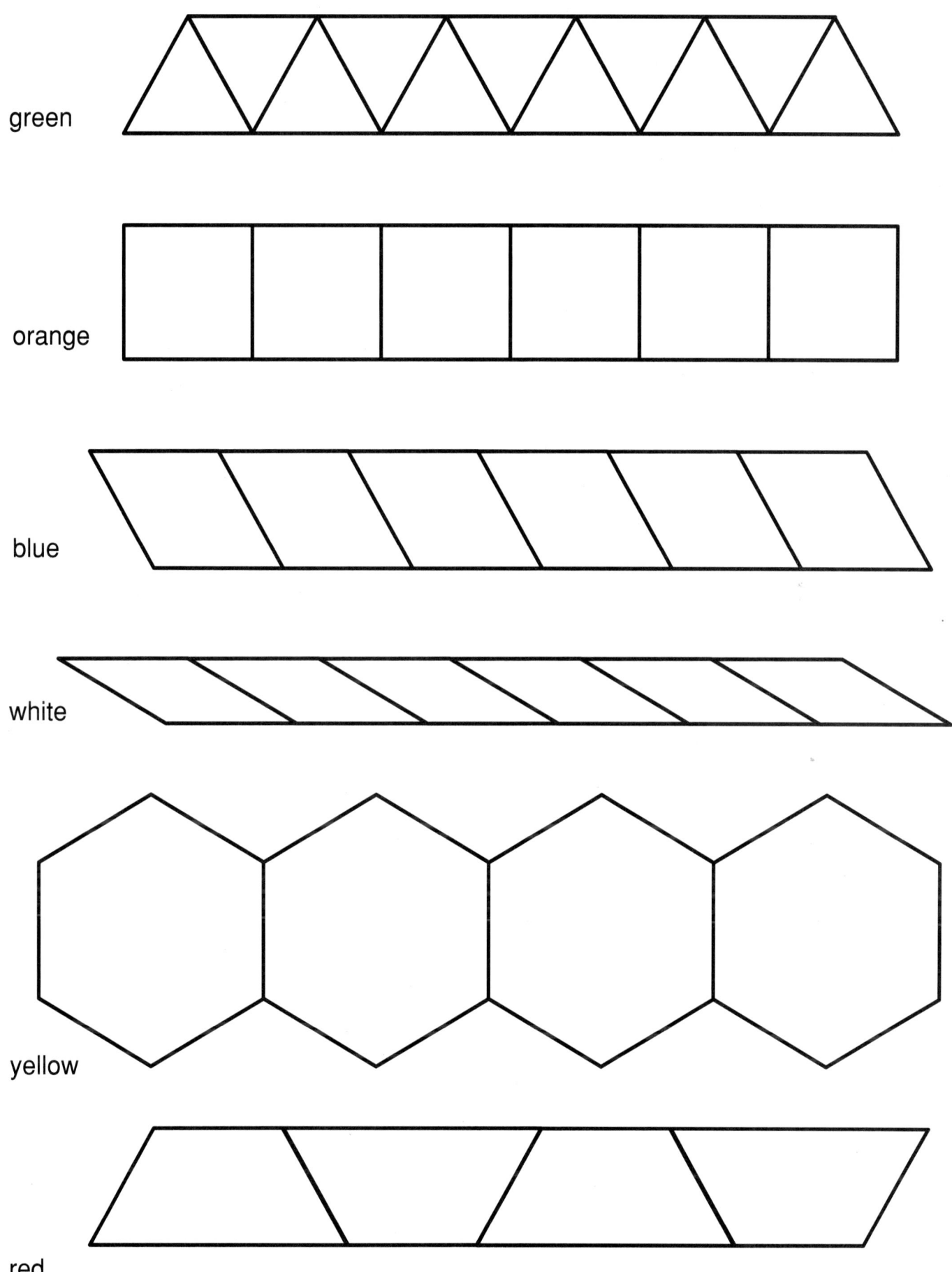

Number Cards

0	1	2	3	4
5	6	7	8	9

Spinner

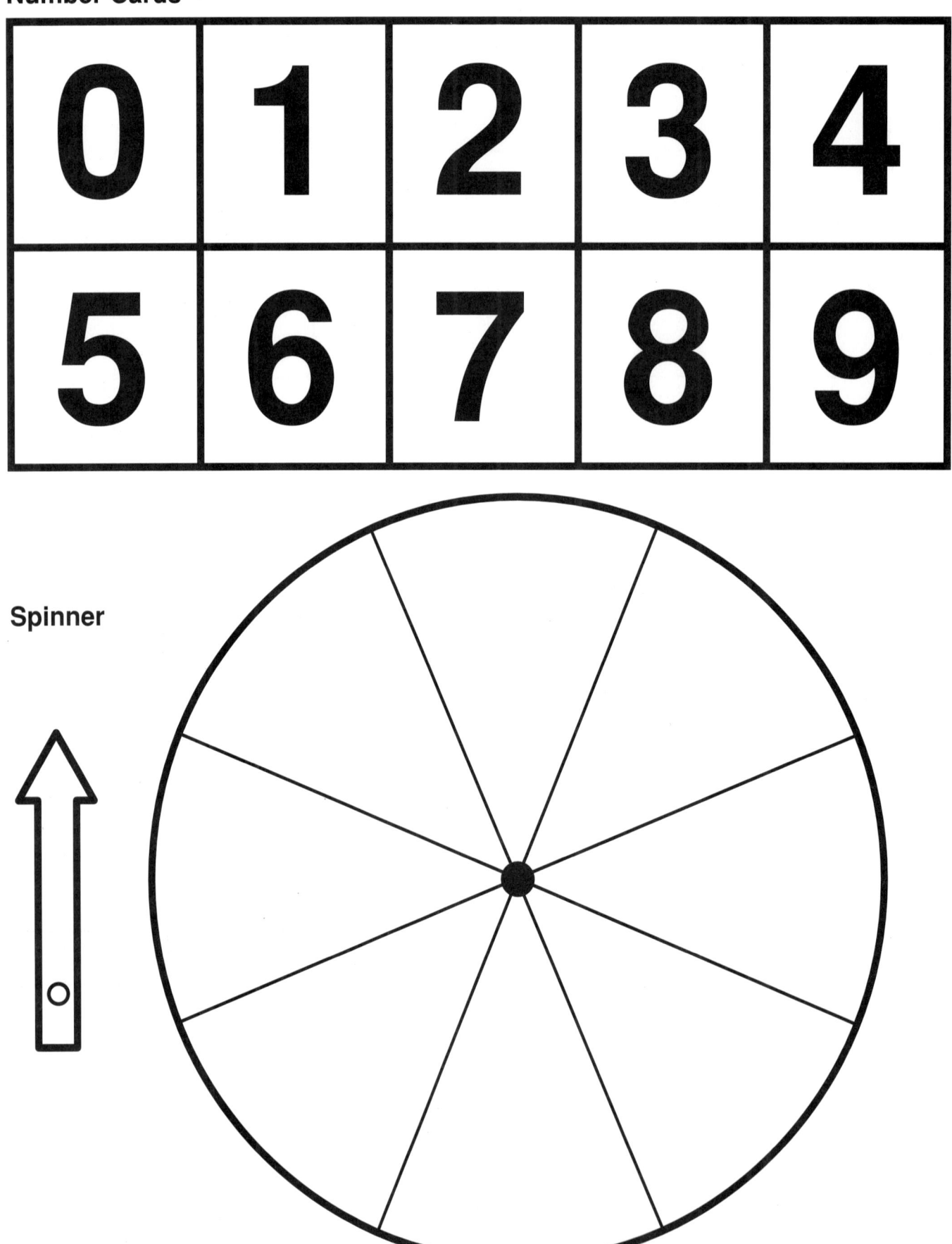

©2001 The Education Center, Inc. • *Math Skills Workout* • TEC3227

Note to the teacher: To assemble the spinner, mount the spinner and pointer to tagboard. Then cut them out and attach the pointer with a metal brad. Or make a construction paper copy of the spinner and have students use a paper clip and pencil instead of the pointer.

Centimeter Graph Paper

One-Inch Graph Paper

Answer Keys

Page 6
1. 19 (cardinal)
2. 4th, 9th, and 14th (ordinal)
3. 5 (cardinal)
4. 1st, 2nd, 6th, 12th, 17th, and 19th (ordinal)
5. 3 (cardinal)
6. 3 (cardinal)
7. 3rd, 5th, 11th, 15th, and 18th (ordinal)
8. 17 (cardinal)
9. 16th (ordinal)
10. 6 (cardinal)
11. 9 (cardinal)
12. 10th (ordinal)

Bonus Box: Answers will vary.

Page 8
1. 63, 36
2. 721
3. 12
4. 192
5. odd, odd
6. 10, even
7. 4, even
8. 15, odd

Bonus Box: Students should have colored balls with the following numbers red: 0, 2, 4, 6, 8. Students should have colored balls with the following numbers yellow: 1, 3, 5, 7, 9.

Page 10
A. 230
B. 126
C. 412
D. 132
E. 301
F. 253
G. 340
H. 404
I. 3 in hundreds ring, 1 in tens ring, 4 in ones ring
J. 2 in hundreds ring, 6 in tens ring, 2 in ones ring
K. 5 in hundreds ring, 0 in ring, 1 in ones ring
L. 1 in hundreds ring, 6 in tens ring, 0 in ones ring

Bonus Box: One arrow must land in the hundreds ring, and the other 2 must land in the ones ring.

Page 12
A. 1,502   B. 8,769   C. 9,012   D. 9,817
E. 1,203   F. 9,087   G. 4,012   H. 9,786

The value of each digit with a star beneath it is as follows.
A. 1,000   B. 700   C. 10   D. 800
E. 3   F. 9,000   G. 2   H. 80

Bonus Box: The largest number is 9,876. The smallest number is 1,023.

Page 14

Order #1,041 yellow	Order #1,178 yellow	Order #1,443 yellow	Order #1,972 yellow	Order #1,999 yellow	Order #2,628 yellow
Order #2,673 yellow	Order #3,117 yellow	Order #3,333 yellow	Order #4,592 orange	Order #4,628 orange	Order #5,000 orange
Order #5,203 orange	Order #6,500 orange	Order #7,280 orange	Order #7,417 orange	Order #7,498 orange	Order #8,429 orange

Bonus Box: Answers will vary.

Page 16
1. 95,657   2. 31,325   3. 40,293   4. 10,300   5. 15,555
6. 50,217   7. 64,177   8. 12,328   9. 79,199   10. 20,412

Bonus Box: Students should have used the code to color the sections on each truck.

Page 18
A. 50
B. 20
C. 100
D. 70
E. 500
F. 600
G. 10
H. 50
I. 200
J. 70
K. 20
L. 90
M. 600
N. 200
O. 500

Bonus Box: Answers will vary.

Page 20
Order of fact families may vary.

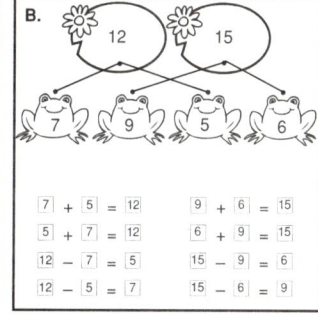
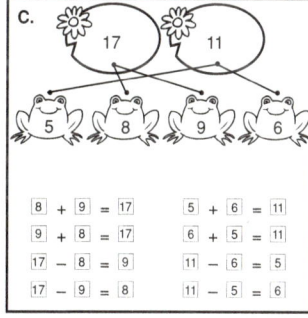
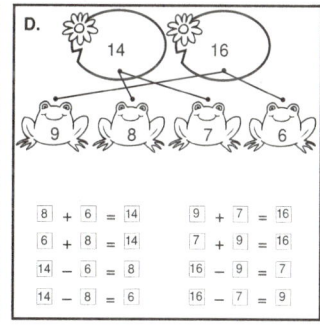

Bonus Box: Answers will vary.

Page 22
A. 54   B. 85   C. 73   D. 60
E. 71   F. 76   G. 51   H. 83
I. 46   J. 72   K. 75   L. 64

Bonus Box: 85, 83, 76, 75, 73, 72, 71, 64, 60, 54, 51, 46

Page 24
A. 66   B. 54   C. 53   D. 63
E. 98   F. 85   G. 81   H. 94
I. 84   J. 76   K. 97   L. 91
M. 73   N. 89   O. 77   P. 90

Bonus Box: Students should have written 3 addends that total 60 and 3 addends that total 83.

Page 26

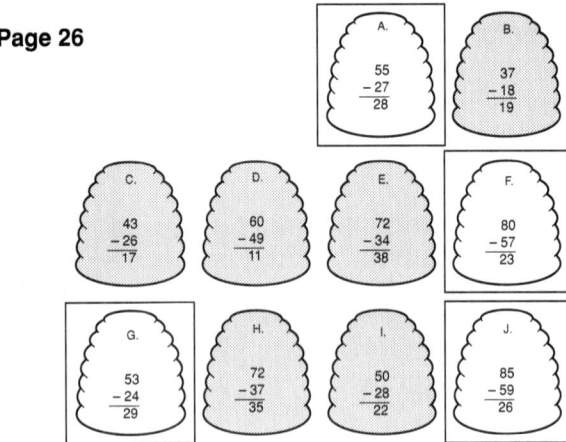

Bonus Box: Answers will vary.

Page 28
1. 148   2. 65   3. 346   4. 294
5. 268   6. 222   7. 28   8. 736
9. 166   10. 377 11. 353 12. 168

Riddle answer: It must be "Wins-day"!
Bonus Box: Answers will vary.

Page 30
1. 2, 10, 2 x 10 = 20
2. 3, 4, 3 x 4 = 12
3. 2, 8, 2 x 8 = 16
4. 4, 5, 4 x 5 = 20
5. 3, 6, 3 x 6 = 18
6. 1, 8, 1 x 8 = 8
7. 5, 3, 5 x 3 = 15
8. 6, 3, 6 x 3 = 18

Bonus Box: Answers may vary. Possible answers include
1. February   2. March
3. April   4. May
5. September   6. October
7. November   8. December
                  or January

Page 32
A. 6 x 3 = 18
B. 3 x 5 = 15
C. 10 x 1 = 10
D. 6 x 4 = 24
E. 5 x 5 = 25
F. 4 x 4 = 16
G. 8 x 3 = 24
H. 10 x 2 = 20
I. 9 x 2 = 18
J. 3 x 7 = 21
K. 12 x 2 = 24
L. 16
M. 9
N. 12
O. 20
P. 12

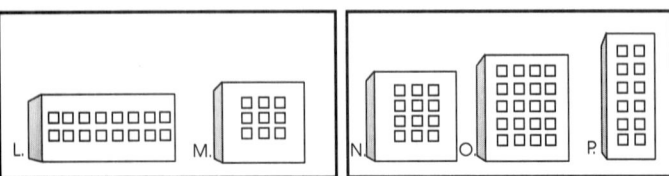

Bonus Box: 2 x 7 = 14, 7 x 2 = 14, 1 x 14 = 14, or 14 x 1 = 14

Page 34

Start With...	Number to Put in Each Group	Number of Groups You Get	Matching Illustration	Matching Division Problem
8 marbles	2	4		8 ÷ 2 = 4
9 marbles	3	3		9 ÷ 3 = 3
10 marbles	2	5		10 ÷ 2 = 5
15 marbles	5	3		15 ÷ 5 = 3
12 marbles	6	2		12 ÷ 6 = 2
16 marbles	4	4		16 ÷ 4 = 4
14 marbles	7	2		14 ÷ 7 = 2
12 marbles	4	3		12 ÷ 4 = 3
16 marbles	8	2		16 ÷ 8 = 2
18 marbles	6	3		18 ÷ 6 = 3

Bonus Box: Students should have illustrated that the marbles could be divided equally into groups of 2, 3, and 6.

Page 36
The order of facts may vary. Sets C and H should be circled.
A. 3 x 6 = 18   E. 6 x 8 = 48
   6 x 3 = 18   8 x 6 = 48
   18 ÷ 6 = 3   48 ÷ 8 = 6
   18 ÷ 3 = 6   48 ÷ 6 = 8
B. 7 x 8 = 56   F. 3 x 4 = 12
   8 x 7 = 56   4 x 3 = 12
   56 ÷ 7 = 8   12 ÷ 3 = 4
   56 ÷ 8 = 7   12 ÷ 4 = 3
D. 2 x 9 = 18   G. 4 x 9 = 36
   9 x 2 = 18   9 x 4 = 36
   18 ÷ 9 = 2   36 ÷ 9 = 4
   18 ÷ 2 = 9   36 ÷ 4 = 9

Bonus Box: In set C, students should have changed 4 to 5 or 35 to 28. In set H, students should have changed 7 to 5 or 15 to 21. Then students should have written the fact family on the corresponding suitcase for each set of numbers.

Page 38

halves	eighths			fourths	thirds	fifths
	thirds	fourths		thirds		ninths
		eighths	fourths	halves		fifths

Bonus Box: Students should have drawn 4 figures. One figure should show halves, 1 should show thirds, 1 should show fourths, and 1 should show sixths.

Page 40
A. $\frac{1}{3}$ = resting   B. $\frac{1}{2}$ = digging
   $\frac{2}{3}$ = not resting $\frac{1}{2}$ = not digging
C. $\frac{3}{4}$ = hopping   D. $\frac{2}{3}$ = eating
   $\frac{1}{4}$ = not hopping $\frac{1}{3}$ = not eating
E. $\frac{2}{5}$ = drinking   F. $\frac{1}{4}$ = babies
   $\frac{3}{5}$ = not drinking $\frac{3}{4}$ = adults

Bonus Box: Students should have drawn 12 kangaroos and colored 6 of them red, 2 of them brown, and 4 of them yellow.

Page 42
A. =
B. >
C. =
D. >
E. <
F. >
G. >
H. <
I. =
J. =
K. <
L. >
M. <

Bonus Box: Each student should have drawn an illustration to show that Flossy ran the farthest because $\frac{3}{4}$ of a mile is greater than $\frac{1}{2}$ and $\frac{2}{3}$ of a mile.

Page 44
The following pairs of spiders should have draglines with the same color: A and E, B and H, C and F, D and G.
I. $\frac{2}{8}$ J. $\frac{2}{4}$ K. $\frac{3}{3}$ L. $\frac{2}{10}$ M. $\frac{2}{12}$ N. $\frac{1}{2}$
Bonus Box: Possible answers include $\frac{2}{4}$, $\frac{3}{6}$, $\frac{4}{8}$, $\frac{5}{10}$, and $\frac{6}{12}$.

Page 46
1. $2\frac{3}{4}$ 2. $2\frac{1}{3}$ 3. $1\frac{1}{6}$
4. $2\frac{3}{6}$ or $2\frac{1}{2}$ 5. $3\frac{3}{8}$ 6. $2\frac{1}{5}$

7.    8. ⊕⊕   9. ⊕⊕⊕
10.   11. ⊕⊕⊕   12. ⊕⊕

Bonus Box: There were 20 slices of pizza left.

Page 48
1. $\frac{9}{10}$ 10. .6
2. .2   11. $\frac{3}{10}$
3. .7   12. $\frac{7}{10}$
4. $\frac{1}{10}$ 13. .9
5. .3   14. .1
6. $\frac{4}{10}$ 15. .4
7. .5   16. $\frac{5}{10}$
8. $\frac{2}{10}$ 17. .8
9. $\frac{8}{10}$ 18. $\frac{6}{10}$

Riddle answer: He wanted to get a hole in one!
Bonus Box: a dime

Page 50
Actual measure for each snake is as follows:
1. 3 candies
2. 9 candies
3. 8 candies
4. 4 candies
5. 8 candies
6. 6 candies
7. 12 candies
8. 5 candies
9. 10 candies
10. 4 candies
Bonus Box: Answers will vary.

Page 52
Estimates will vary. Actual measurement for each leap is as follows:
1. $4\frac{1}{2}$ inches
2. 8 inches
3. $2\frac{1}{2}$ inches
4. 6 inches
5. 7 inches
6. $8\frac{1}{2}$ inches
7. $4\frac{1}{2}$ inches
8. 3 inches
9. $5\frac{1}{2}$ inches
10. $4\frac{1}{2}$ inches
Bonus Box: 21 inches

Page 54
Estimates for 1–6 will vary. Each actual measurement is below.
1. 8 cm
2. 4 cm
3. 6 cm
4. 3 cm
5. 9 cm
6. 10 cm
7. meters
8. meters
9. centimeters
10. meters
11. centimeters
12. centimeters
Bonus Box: Answers will vary.

Page 56
1. ounces
2. ounces
3. pounds
4. ounces
5. pounds
6. ounces
7. ounces
8. ounces
9. ounces
10. pounds
11. pounds
12. ounces
13. ounces
14. pounds
15. pounds
Bonus Box: Answers will vary.

Page 58
1. X
2. ✔
3. X
4. X
5. ✔
6. ✔
7. X
8. X
9. ✔
10. ✔
11. ✔
12. ✔
13. X
14. X
Bonus Box: Answers will vary.

Page 60
Answers are as follows. However, accept any other reasonable responses for A–L.
A. red   J. green
B. green   K. orange
C. red   L. green
D. orange   M. 2 pints
E. green   N. 3 pints
F. yellow   O. 18 cups
G. yellow   P. 1 quart
H. green   Q. 2 pints
I. orange   R. 1 gallon
Bonus Box: cups = 8, pints = 4, quarts = 2

Page 62
Items listed for A–F should be circled on each suitcase.
A. shorts, tank top
B. sandals, T-shirt
C. swimsuit, tank top
D. shorts, T-shirt
E. coat, gloves
F. coat, scarf
G. 72°F, fishing
H. 24°F, ice-skating
I. 58°F, hiking
J. 98°F, swimming
Bonus Box: Answers will vary.

Page 64
A. 32°C
B. 15°C
C. 28°C
D. 40°C
E. −23°C
F. 7°C
G. −30°C
H. 37°C
I. The thermometer should be shaded and labeled to show −15°C. *Sledding* should be listed as the corresponding activity.
J. The thermometer should be shaded and labeled to show 25°C. *Gardening* should be listed as the corresponding activity.
K. The thermometer should be shaded and labeled to show 35°C. *Swimming* should be listed as the corresponding activity.
Bonus Box: Answers will vary.

Page 66

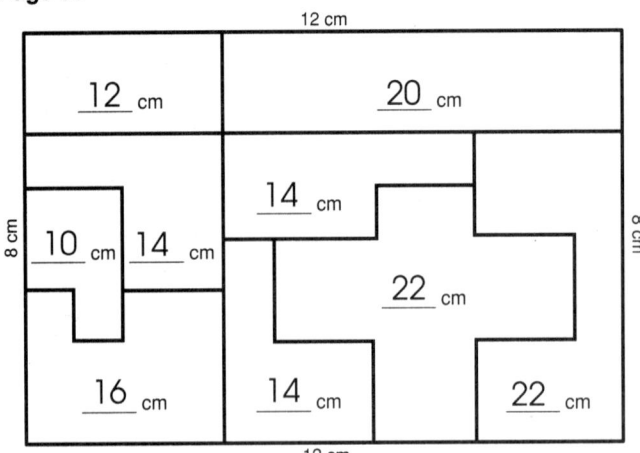

Bonus Box: 40 cm

Page 68

The length and width in each row can be reversed. However, the area remains the same.

1. 30 sq. cm
2. 144 sq. cm
3. The area is equal to length times width.

Activity Zone	Length	Width	Area
Skating Rink	3 cm	3 cm	9 sq. cm
Baseball Field	3 cm	6 cm	18 sq. cm
Miniature Golf Course	6 cm	4 cm	24 sq. cm
Batting Cage	1 cm	5 cm	5 sq. cm
Tennis Court	3 cm	5 cm	15 sq. cm
Pool	5 cm	2 cm	10 sq. cm
Soccer Field	3 cm	7 cm	21 sq. cm
Volleyball Court	2 cm	7 cm	14 sq. cm
Bowling Alley	2 cm	4 cm	8 sq. cm
Basketball Court	4 cm	5 cm	20 sq. cm

Bonus Box:

Page 70

1. 6 cubic units
2. 9 cubic units
3. 8 cubic units
4. 10 cubic units
5. 24 cubic units
6. 12 cubic units
7. 16 cubic units
8. 15 cubic units
9. 24 cubic units
10. 27 cubic units

Bonus Box: Students should have circled figures 5, 7, and 10.

Page 72

A. 8:45 (green)
B. 5:00 (red)
C. 4:30 (blue)
D. 9:15 (yellow)
E. 11:30 (blue)
F. 1:45 (green)
G. 12:00 (red)
H. 7:15 (yellow)
I. 3:30 (blue)
J. 3:45 (green)
K. 6:00 (red)
L. 10:15 (yellow)

Bonus Box: Answers will vary.

Page 74

Each clock should show the time indicated below.

A. 8:10
B. 8:40
C. 9:20
D. 9:25
E. 9:50
F. 10:15
G. 10:35
H. 10:55
I. 11:05
J. 11:45

Bonus Box: Answers will vary.

Page 76

A. 7:28—Make muffins.
B. 1:17—Make fruit pies.
C. 4:52—Clean and mop kitchen.
D. 11:25—Take and fill orders.
E. 9:48—Bake cookies and cakes.
F. 1:03—Make fruit pies.
G. 10:24—Bake cookies and cakes.
H. 12:07—Take a lunch break.
I. 9:27—Make dough for sweet rolls.
J. 11:05—Take and fill orders.

Bonus Box: Answers will vary.

Page 78

1. 20 minutes
2. 15 minutes
3. 40 minutes
4. 1 hour and 20 minutes
5. 1 hour and 15 minutes
6. 45 minutes
7. 30 minutes
8. 20 minutes
9. 25 minutes
10. 1 hour and 25 minutes

Bonus Box: 11 hours and 25 minutes

Page 80

1. left safe: 83¢
   right safe: 73¢
   The left safe should be colored.
2. left safe: 89¢
   right safe: 76¢
   The left safe should be colored.
3. left safe: 69¢
   right safe: 74¢
   The right safe should be colored.
4. left safe: 98¢
   right safe: 95¢
   The left safe should be colored.
5. left safe: 42¢
   right safe: 40¢
   The left safe should be colored.

Bonus Box: Answers will vary.

Page 82

The cost and total change for each item is shown. The coins marked on each student's chart will vary.

1. $0.49, $0.51
2. $2.10, $0.90
3. $0.77, $0.23
4. $0.60, $0.40
5. $1.70, $0.30
6. $1.23, $0.77
7. $0.37, $0.63
8. $0.85, $0.15
9. $0.28, $0.72
10. $2.20, $0.80

Page 84

1. $10.00
2. $0.51
3. $12.50
4. $2.00
5. $2.00
6. $11.75
7. $3.50
8. $7.00
9. $3.00
10. $2.73

Bonus Box: The puppy and supplies cost the most. The goldfish and supplies cost the least.

Page 86

1. octagon
2. triangle
3. square
4. pentagon
5. rectangle
6. hexagon
7. parallelogram
8. trapezoid

Bonus Box: Possible answers include a square, rectangle, parallelogram, rhombus, and trapezoid.

Page 88
(Order of polygons will vary in each row.)

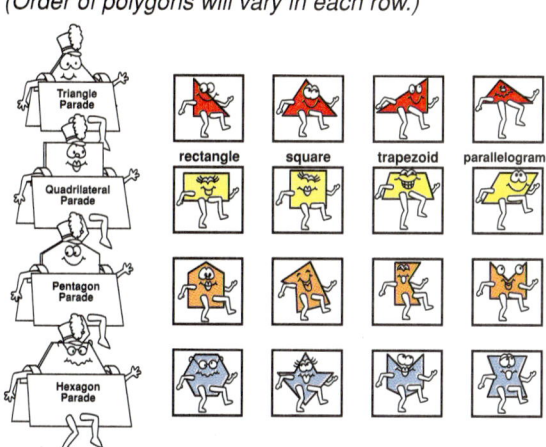

Bonus Box: See labels above.

Page 90
1. rectangular prism
2. pyramid
3. cylinder
4. sphere
5. rectangular prism
6. cylinder
7. rectangular prism
8. cone
9. cube
10. cylinder
11. cone
12. rectangular prism
13. cube
14. pyramid
15. sphere

Bonus Box: Answers will vary.

Page 92
1. cone 1, 0, 0
2. rectangular prism 6, 12, 8
3. sphere 0, 0, 0
4. cube 6, 12, 8
5. pyramid 5, 8, 5
6. cylinder 2, 2, 0
7. sphere and cone
8. cone and cylinder
9. pyramid and cube
10. rectangular prism and pyramid

Bonus Box: Answers will vary.

Page 94
1. ray LM
2. line BC
3. ray JK
4. line segment YZ
5. line segment EF
6. ray OP
7. line segment GH
8. line QR

Bonus Box: Answers will vary.

Page 96
The Cup
1. 1+
2. 1
3. 0
4. 1
5. 1
6. 1

The Whale
1. 1+
2. 1
3. 1
4. 1
5. 1
6. 0
7. 0
8. 0
9. 0

Page 98

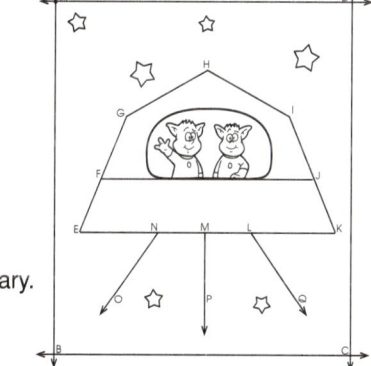

Bonus Box: Answers will vary.

Page 100
(Order of angles in each squad will vary.)

Bonus Box: Answers will vary.

Page 102
1. flip
2. turn
3. slide
4. turn
5. slide
6. flip
7. turn
8. slide
9. flip

Bonus Box: Answers will vary.

Page 104
Route #1
1. (6, 4)
2. (1, 2)
3. (2, 5)
4. (4, 2)
5. (7, 0)
6. (3, 4)
7. (7, 2)
8. (0, 4)

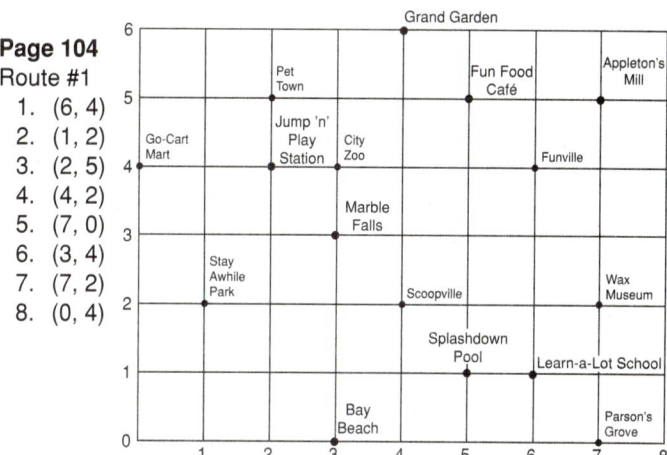

Route #2: Answers should be written on grid as shown.
Bonus Box: Answers will vary.

Page 106

12 green triangles	6 red trapezoids	5 orange squares
4 blue parallelograms / 4 tan rhombuses	6 red trapezoids / 1 blue parallelogram	4 green triangles / 4 orange squares
11 pattern blocks / 8 green triangles / 2 orange squares / 1 tan rhombus	9 pattern blocks / 1 orange square / 4 red trapezoids / 4 tan rhombuses	7 pattern blocks / 4 tan rhombuses / 2 red trapezoids / 1 blue parallelogram

Bonus Box: Answers will vary.

Page 108
1. 16
2. 9
3. 15
4. 40
5. 6
6. Winner's Whirl, Mighty Tigers, Victory Growl
7. 4
8. 2

| Tally Table — Player's Favorite Cheers ||
Cheer	Tallies
Victory Growl	||||| ||||| |||||	
Winner's Whirl	||||| ||||
Mighty Tigers	||||| ||||| |||||

Bonus Box: Students should have completed a tally table similar to the one shown.

Words	Tallies
tigers	|||||
rhyme with *eat*	||
rhyme with *owl*	||

Page 110
1. vanilla
2. milk
3. punch
4. 1:00 P.M.
5. Possible explanation: Her friends might have chosen a bigger variety of flavors, drinks, and times.

Bonus Box: Answers will vary.

Ice-Cream Flavor	Tallies
Chocolate	||||
Vanilla	|||||	
Strawberry	||

Drink	Tallies
Soda	||||
Milk	|||||
Punch	|||

Time	Tallies
1:00 P.M.	|||||	
3:00 P.M.	||
7:00 P.M.	||||

Page 112
1. Tasty Tortillas, 11 grams
2. Presto Popcorn, 5 grams
3. Chipper Chips and Mega Munchies
4. 5 grams
5. 14 grams
6. grams of carbohydrates in Chipper Chips
7. Mega Munchies
8. Presto Popcorn

Bonus Box:

| Calories in Snacks ||
Snacks	Calories
Chipper Chips	11 symbols
Presto Popcorn	7 symbols
Tasty Tortillas	13 symbols
Mega Munchies	9 symbols
Key: ◐ = 10 calories	

Page 114
1. Answers will vary.
2. news
3. cooking
4. sports and movies
5. 2½ hours

Bonus Box: Answers will vary.

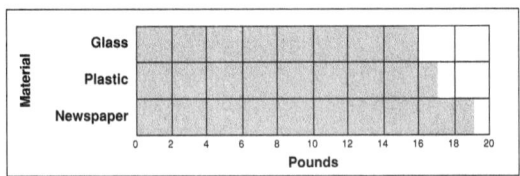

Page 116
1. higher
2. Adrian
3. Brett
4. 15
5. 20

Bonus Box: Answers will vary.

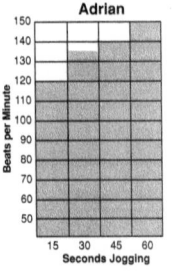 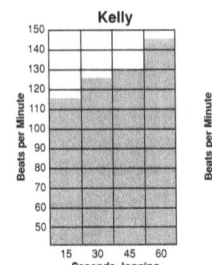 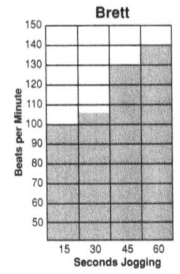

Page 118
1. newspaper
2. 52 pounds
3. 3 pounds
4. 33 pounds
5. Pounds of Material Collected
6. How many students collected glass?

Bonus Box: Day 1—19 pounds, Day 2—13 pounds, Day 3—20 pounds

Page 120
1. shark, sea horse
2. starfish and jellyfish
3. Answers will vary.
4. Answers will vary.
5. Possible explanation: Each animal would have been picked a relatively equal number of times.

Bonus Box: all starfish cards, all except shark cards

Page 122
1. hot dogs, chicken, sandwiches
2. hamburgers, since this section represents the largest part of the spinner
3. Answers will vary.
4. Answers will vary.
5. Answers will vary.

Bonus Box: Answers will vary.

Page 124
1. 6
2. 2
3. 4
4. a. unlikely
   b. likely
   c. certain
   d. impossible
   e. equally likely
5.

Results				
14	40	33	33	23
27	33	40	18	40
18	23	18	40	27
33	18	27	27	18

| Frequency Table ||
Number	Frequency
less than 20	6
greater than 20	14

6. Answers will vary. Accept all reasonable explanations.

Bonus Box: Answers will vary.

Page 126
1. 2, 3, 4, 5, or 6
2. 3, 3
3. Yes. The probability of rolling an odd or even number is equally likely. Therefore, each player has an equal chance of winning.
4. No. There is not an equal number of even and odd numbers. Therefore, each player does not have an equal chance of winning.
5. Answers will vary.

Bonus Box: Answers will vary. However, the unfair die for Player 1 should include more even than odd numbers. The unfair die for Player 2 should include more odd than even numbers.

Page 128
Students should have described the patterns in pictures 1, 4, 5, and 6. Possible answers are shown below. Students should have circled pictures 2, 3, 7, and 8.
1. The pattern shows a penguin standing, then a penguin lying down.
4. The pattern shows 1 big fish, then 2 small fish.
5. The pattern shows squares, then alternating lines and zigzags.
6. The pattern shows 1 large snowflake, then 1 small snowflake.

Bonus Box: Answers will vary.

Page 130

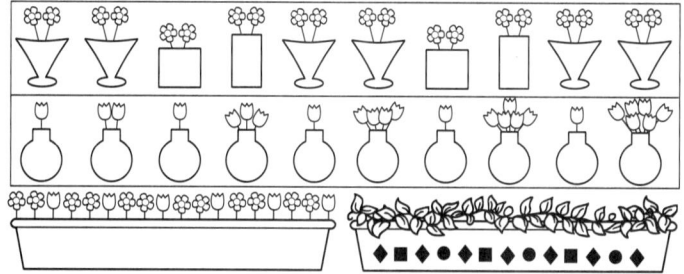

Bonus Box: orange

Page 132
A. 9 B. 6 C. 8 D. 7
E. 5 F. 6 G. 1 H. 4
I. 8 J. 7 K. 9 L. 7
M. 10 N. 9 O. 5 P. 2
Q. 4 R. 8 S. 7 T. 7

Bonus Box: One possible answer: Yes. To find the missing addend, you can subtract the given addend from the sum.

Page 134
A. 3   H. 1
B. 8   I. 4
C. 2   J. 0
D. 6   K. 4
E. 5   L. 6
F. 9   M. 7
G. 7   N. 6

Bonus Box: Ice-cream treats D and K should be circled.

Page 136
A. 28   B. 9   C. 35
   7 x 4 = 28   3 + 6 = 9   7 x 5 = 35

D. 24   E. 5   F. 1
   8 x 3 = 24   2 + 3 = 5

G. 15   H. 3   I. 4
   8 + 7 = 15

J. 10   K. 36   L. 7
   2 x 5 = 10   4 x 9 = 36   2 + 5 = 7

1. No. Explanations will vary.
2. Yes. Explanations will vary.
3. Yes. Explanations will vary.

Bonus Box: Answers will vary.

Page 138
A. 21   H. 6   O. 4 x (7 − 3) = 16
B. 28   I. 0   P. (3 + 9) x 3 = 36
C. 56   J. 4   Q. 4 + (4 x 3) = 16
D. 8   K. 12   R. (9 + 1) x 3 = 30
E. 18   L. 14   S. 7 x (4 − 1) = 21
F. 20   M. (5 − 3) x 2 = 4   T. 7 x (1 + 3) = 28
G. 50   N. (2 x 6) + 5 = 17 U. 4 + (3 x 2) = 10

Bonus Box: (4 x 1) + 5 = 9, 4 x (1 + 5) = 24

Page 140

1. Code: A

Input	Output
10	7
14	11
6	3
9	6
5	2

2. Code: B

Input	Output
3	10
5	12
14	21
21	28
0	7

3. Code: G

Input	Output
3	15
6	30
8	40
2	10
10	50

4. Code: E

Input	Output
8	16
10	20
4	8
15	30
50	100

5. Code: R

Input	Output
5	20
15	30
20	35
30	45
10	25

6. Code: L

Input	Output
26	14
15	3
13	1
20	8
12	0

An A L G E B R A activity!
   1 6 3 4 2 5 1

Bonus Box: Answers will vary.

Page 142
A. 8   G. 16
B. 15   H. 6
C. 12   I. 13
D. 10   J. 17
E. 7   K. 5
F. 9   L. 5

Bonus Box: 9 birds (☐ + 2 − 5 = 6; 6 + 5 − 2 = 9)

Page 144

Bonus Box: One possible explanation: The tomato plants must be in the second and third columns because the pepper plants are in the first and fourth columns.

Page 146
1. 6   5. 2
2. 2   6. Freddy
3. Brian   7. 5
4. 3

Bonus Box: Answers will vary.

Page 148

The position of flowers on the lei will vary. However, there should be 3 blue flowers, 5 yellow flowers, 2 red flowers, 4 orange flowers, and 1 purple flower.

1. yellow
2. 7
3. 4
4. 8
5. 15

Bonus Box: 7

Page 150

Students' sentences will vary. Possible sentences include the following:

1. Two of the cars did not get tires.

Cars	1	2	3	4	5	6	7	8	9	10	11	12
Tires	4	8	12	16	20	24	28	32	36	40	X	X

2. There were 21 bags of peanuts sold.

Hot Dogs	7	14	21	28	35	42	49
Peanuts	3	6	9	12	15	18	21

3. The drivers waved in the same lap 2 times.

Laps	1	2	3	4	5	6	7	8	9	10	11	12
Driver 12		X		X		X		X		X		X
Driver 30			X		X			X			X	

4. After 12 laps, 27 cars got gas.

Laps	2	4	6	8	10	12
Cars	2	5	9	14	20	27

5. Three children met the drivers.

Fans	1	2	3	4	5	6	7	8	9	10	11	12	13	14	15
Adult	X	X	X		X	X	X		X	X	X		X	X	X
Child				X				X				X			

Bonus Box: Five of the fans would be children.

Page 152

Students should have drawn symbols on the calendar to show that Rita's next…

- game is October 25.
- scrimmage is October 22.
- 2 practices are October 26 and 31.

1. 11, 16, 21, 26, 31
   The rule for the pattern is add 5.
2. 8, 18, 22
   The rule for the pattern is add 4, then 10.
3. 9, 17, 25
   The rule for the pattern is add 8.
4. Thursday
5. November 2, November 1, November 5

Bonus Box: October 25

Page 154

1. 6 choices
   Side Street Boyz CD   Side Street Boyz cassette
   Jenny Mopeds CD   Jenny Mopeds cassette
   N'Line CD   N'Line cassette

2. 12 choices
   16" gold, heart   18" gold, heart   16" silver, heart   18" silver, heart
   16" gold, star   18" gold, star   16" silver, star   18" silver, star
   16" gold, flower   18" gold, flower   16" silver, flower   18" silver, flower

3. 9 choices
   sandals, red   clogs, red   sneakers, red
   sandals, blue   clogs, blue   sneakers, blue
   sandals, yellow   clogs, yellow   sneakers, yellow

4. 4 choices
   orange shirt, tan pants   green shirt, tan pants
   orange shirt, black pants   green shirt, black pants

5. 6 choices
   chocolate, peanut butter   peanut butter, maple
   chocolate, maple   peanut butter, vanilla
   chocolate, vanilla   maple, vanilla

Bonus Box: Answers will vary.

Page 156

Some of the following have more than 1 answer. A possible answer for each is shown below.

1. Sam: 8 ride, 5 game, and 5 food tickets
   Juan: 10 ride, 3 game, and 4 food tickets
2. Sam: 2, 6, 8, 12
   Juan: 4, 6, 8, 12
3. Sam: 4
   Juan: 8
4. Sam: jacks, jump rope, glow ball
   Juan: beanbag, whistle, yo-yo, jump rope
5. Sam: burger, nachos, soda
   Juan: hot dog, salad, water

Bonus Box: Students should have underlined the first sentence in problem 2.

Page 158

1. red or yellow, 20 runners
2. red, 1 hour and 30 minutes
3. green, 3 races
4. blue, 4 more sprinters
5. red or yellow, 40 miles
6. blue, 32 miles
7. red, 43 races
8. blue, 23 meets
9. green, 6 runners
10. red or yellow, 72 uniforms

Bonus Box: Each runner would have run 2 races each.

Page 160

Strategies may vary. Possible strategies are listed below.

1. act it out, elephant
2. make a list, 6 combinations
3. guess and check, 22 minutes, 12 minutes
4. make a table, 15 boxes
5. find a pattern, 9 acrobats
6. use logical reasoning, Kim—popcorn, Tim—cotton candy, Jim—candied apple

Bonus Box: The student should have chosen the draw-a-picture strategy to solve problem 6.